Walter Glöckle

The Quantum Mechanical Few-Body Problem

With 17 Figures

Springer-Verlag
Berlin Heidelberg New York Tokyo 1983

Professor Dr. Walter Glöckle

Institut für Theoretische Physik, Ruhr-Universität Bochum
D-4630 Bochum 1, Fed. Rep. of Germany

Editors

Wolf Beiglböck

Institut für Angewandte Mathematik
Universität Heidelberg
Im Neuenheimer Feld 5
D-6900 Heidelberg 1, Fed. Rep. of Germany

Elliott H. Lieb

Department of Physics
Joseph Henry Laboratories
Princeton University
Princeton, NJ 08540, USA

Tullio Regge

Istituto di Fisica Teorica
Università di Torino, C. so M. d'Azeglio, 46
I-10125 Torino, Italy

Walter Thirring

Institut für Theoretische Physik
der Universität Wien, Boltzmanngasse 5
A-1090 Wien, Austria

ISBN-13:978-3-642-82083-0   e-ISBN-13:978-3-642-82081-6
DOI: 10.1007/978-3-642-82081-6

Library of Congress Cataloging in Publication Data. Glöckle, Walter, 1939–. The quantum mechanical few-body problem. (Texts and monographs in physics). Bibliography: p. Includes index. 1. Few-body problem. 2. Quantum theory. I. Title. II. Series. QC174.17.P7G58   1983   530.1′2   83-10325

Typesetting: K + V Fotosatz, Beerfelden.

2153/3130-543210

Preface

Few-body systems are both technically relatively simple and physically non-trivial enough to test theories quantitatively. For instance the He-atom played historically an important role in verifying predictions of QED. A similar role is contributed nowadays to the three-nucleon system as a testing ground for nuclear dynamics and maybe in the near future to few-quark systems. They are also often the basic building blocks for many-body systems like to some extent nuclei, where the *real many-body* aspect is not the dominant feature.

The presentation of the subject given here is based on lectures held at various places in the last ten years. The selection of the topics is certainly subjective and influenced by my own research interests. The content of the book is simply organized according to the increasing number of particles treated. Because of its conceptual simplicity single particle motion is very suitable for introducing the basic elements of scattering theory. Using these elements the two-body system is treated for the specific case of two nucleons, which is of great importance in the study of the nuclear interaction. Great space is devoted to the less trivial few-body system consisting of three particles. Again physical examples are taken solely from nuclear physics. Finally the four-particle system is discussed so as to familiarize the reader with the techniques required for the formulations of n-bodies in general. One of the aims of the n-body connected kernel formulations is to put conventional, intuitively invented nuclear models and reaction theories on a firm basis. Though there are already promising insights available, the break-through has apparently not yet been found and the natural and desired extension of the matter developed here is still on the "second sheet".

In order not to overload the content of these introductory notes and partially because of existing presentations certain techniques and subjects are not dealt with. These are variational methods, the use of hyperspherical harmonics, the elaboration of finite rank approximations of t-operators and kernels (which played and still play an important role), the very interesting problem of formulating a relativistic theory for n particles, and the whole dynamical problem of nuclear forces which includes the very successul recent solution of few-body Bethe-Salpeter equations.

In the techniques and subject treated there exists a large amount of publications. We would like to apologize to those authors whose work is not directly or sufficiently well mentioned. There are several reviews and articles related to our subject. Besides special monographs on few-body systems we refer the reader also to some books which are closely related. An important source of information are the proceedings of the international few-body conferences held up to now. All these sources are cited at the end.

The book is written for students and does not require more than a basic course in QM. It emphasizes also the practical points of view and will hopefully be profitable to some researchers working in that field as well.

This work would not have been undertaken without the continuous stimulation by Professor Hélio T. Coelho. I am very thankful to him and for his kind hospitality which he extended to me at his institute in Recife, where parts of the notes have been written. Dr. R. Brandenburg eradicated my major blunders in English and helped me in some parts to clarify the presentation, for which I thank him very much. Last but not least I want to thank Mrs. Kächele and Mrs. Walter, for their skill and patience in transcribing successfully my handwriting into a legible form.

Bochum, January 1983   *W. Glöckle*

Contents

1. Elements of Potential Scattering Theory

Scattering of a particle by a potential is a simple physical picture but rich enough to introduce such basic concepts of scattering theory as Möller wave operators, in- and outgoing particle flux, unitarity, S-, T- and K-matrices, Lippmann-Schwinger equations, S-matrix pole trajectories, criteria for convergence or divergence of Neumann series, etc. Therefore the first chapter is basic and following ones use the language developed here, while enriching and extending it according to the increase of possible physical processes for two and more particles.

1.1 The Möller Wave Operator

Let us regard the scattering of a particle by a potential. We assume that the potential drops towards zero outside a certain bounded domain D in space. Initially while approaching D, the particle moves freely with a certain momentum. As it crosses D it will experience a force which classically would bend the initial straight line trajectory. Having left D the particle again moves freely but with a final momentum which can be different from the initial one. It will be the task of Chap. 1 to develop techniques for answering the question of how to find the probability for the change in momentum induced by the potential.

To describe the initial state of free motion outside D we have to localize the particle. Let us choose a wave packet $\psi_0(x, t)$, which obeys the time dependent Schrödinger equation

$$H_0\psi_0(x, t) = i \frac{\partial \psi_0(x, t)}{\partial t} \tag{1.1}$$

with the free Hamilton operator

$$H_0 = -\frac{1}{2m} \nabla^2. \tag{1.2}$$

We put $\hbar = c = 1$. Then units for energy and length convenient for nuclear physics result from $\hbar c = 197.33$ MeV fm.

Clearly $\psi_0(x, t)$ will be of the general form

$$\psi_0(x, t) = \frac{1}{(2\pi)^{3/2}} \int dq \, \exp[i(qx - E_q t)] f_0(q),$$ (1.3)

which is a superposition of momentum eigenstates

$$\psi_q^0(x) = \frac{1}{(2\pi)^{3/2}} e^{iqx}$$ (1.4)

with the energies $E_q = q^2/2m$. In a scattering process the momentum distribution $f_0(q)$ will be peaked at an initial momentum q_i.

For example regard

$$f_0(q) = \frac{1}{b^{3/2}} \left(\frac{2}{\pi}\right)^{3/4} \exp[-(q - q_i)^2/b^2].$$ (1.5a)

The quantity b measures the momentum distribution in the beam. It is a simple exercise to evaluate in that case the integral (1.3). The result is

$$\psi_0(x, t) = \frac{1}{(2\pi)^{3/2}} \exp[i(q_i x - E_{q_i} t)]$$

$$\times (2\pi)^{3/4} b^{3/2} \frac{\exp\left[-\frac{b^2}{4} \frac{\left(x - t\frac{q_i}{m}\right)^2}{1 + itb^2/2m}\right]}{\left(1 + \frac{it}{2m} b^2\right)^{3/2}}.$$ (1.5b)

Thus we find a plane wave with the central momentum q_i in a region of space of extension $d \sim b^{-1}$. The center of the wave packet travels along the classical path. The spreading of the wave packet is controlled by the parameter

$$\xi = \frac{t}{2m} b^2 = \frac{L}{2q} b^2 \approx \frac{L}{d} \frac{b}{q}.$$ (1.5c)

Here we introduced a typical length L between source and detector and the average momentum q of the particle. Under ordinary conditions $\xi \ll 1$ and the spreading is negligible.

As the wave packet approaches D it will feel the potential V and its evolution in time will be governed by the time-dependent Schrödinger equation

$$H \Psi(x, t) = i \frac{\partial \Psi(x, t)}{\partial t} \tag{1.6}$$

with the full Hamilton operator

$$H = H_0 + V. \tag{1.7}$$

So we face the question, how is $\Psi(x, t)$ linked to $\psi_0(x, t)$ or in other words how can we select out of the many solutions of (1.6) that specific one which develops out of the initial state $\psi_0(x, t)$? A first guess could be to fix Ψ through

$$\Psi(x, t) \rightarrow \psi_0(x, t) \quad \text{for} \quad t \rightarrow -\infty.$$

This requirement however is too weak, since both wave functions tend point-wise towards zero in that limit, and one cannot distinguish between different initial states ψ_0. In the example (1.5b) ψ_0 tends towards zero pointwise like $|t|^{-3/2}$. This is true in general.

Exercise: Prove that

$$F(t) = \int_0^\infty dq \, q^2 e^{-iq^2 t} f(q)$$

tends towards

$$\text{const}/|t|^{3/2} \quad \text{for} \quad |t| \rightarrow \infty \quad \text{if} \quad f(0) \neq 0.$$

Hint: use the method of steepest descend [1.1].

Although the wave functions spread out with time, leading to smaller and smaller amplitudes at each point x, their norms

$$\|\Psi(t)\| = \sqrt{\int dx \, |\Psi(x, t)|^2} \tag{1.8}$$

are time independent. Therefore, in order to enforce the equality of Ψ and ψ_0 before the particle reaches D we might require

$$\lim_{t \rightarrow -\infty} \|\Psi(t) - \psi_0(t)\| \rightarrow 0. \tag{1.9}$$

Then the question becomes, is (1.9) compatible with the time dependent Schrödinger equations (1.1, 6)?
Equations (1.1, 6) tell us

$$\begin{aligned}|\psi_0(t)\rangle &= \exp[-iH_0(t - t_0)] |\psi_0(t_0)\rangle \\ |\Psi(t)\rangle &= \exp[-iH(t - t_0)] |\Psi(t_0)\rangle \end{aligned} \tag{1.10}$$

and we can write (1.9) as

$$\| \Psi(t) - \psi_0(t) \| = \| \exp[-iH(t-t_0)] \, \Psi(t_0) - \exp[-iH_0(t-t_0)] \, \psi_0(t_0) \|$$
$$= \| \Psi(t_0) - \exp[iH(t-t_0)] \exp[-iH_0(t-t_0)] \, \psi_0(t_0) \|. \tag{1.9a}$$

The second equality follows from the unitarity of $\exp[-iH(t-t_0)]$. Thus the requirement (1.9), together with the time evolution expressed through the Schrödinger equation, will be

$$| \Psi(t_0) \rangle = \lim_{\tau \to -\infty} e^{iH\tau} e^{-iH_0\tau} | \psi_0(t_0) \rangle. \tag{1.11}$$

If that limit exists, then (1.11) is a link between $| \Psi \rangle$ and $| \psi_0 \rangle$, compatible with the Schrödinger equation. Moreover it gives us a prescription for constructing a specific scattering state at the arbitrary time $t = t_0$ which belongs to a certain choice of initial conditions in the infinite past.

The limit in (1.11) defines the Möller wave operator [1.2]

$$\Omega^{(+)} = \lim_{\tau \to -\infty} (e^{iH\tau} e^{-iH_0\tau}) \tag{1.12}$$

and (1.11) reads for an arbitrary time t

$$| \Psi(t) \rangle = \Omega^{(+)} | \psi_0(t) \rangle. \tag{1.13}$$

This relation (1.13) is the formal solution of the scattering problem to a specific choice of initial conditions.

Let us now sketch a proof [1.3] for the existence of $\Omega^{(+)}$. The ensemble of wave packets $\psi_0(x, t)$ (t fixed) defines the space accessible to the particle. For square integrable momentum distributions they span a Hilbert space. Thus we have to show that $\Omega^{(+)}$ exists on the whole Hilbert space. Define

$$W(t) = e^{iHt} e^{-iH_0t} \tag{1.14}$$

and regard

$$\|(W(t_2) - W(t_1)) \, \psi_0(0) \| = \left\| \int_{t_1}^{t_2} dt \, \frac{d}{dt} \, W(t) \, \psi_0(0) \right\|. \tag{1.14a}$$

The limit (1.12) exists if (1.14a) can be shown to be arbitrarily small if $t_1 < t_2 < 0$ and $|t_2|$ is sufficiently large. Now together with the property of W, namely

$$\frac{d}{dt} \, W(t) = i e^{iHt} (H - H_0) \, e^{-iH_0t} = i e^{iHt} V e^{-iH_0t}, \tag{1.15}$$

we can estimate the rhs as

$$\left\| \int_{t_1}^{t_2} dt\, \frac{d}{dt}\, W(t)\, \psi_0(0) \right\| \leqslant \int_{t_1}^{t_2} dt\, \left\| \frac{d}{dt}\, W(t)\, \psi_0(0) \right\|$$

$$= \int_{t_1}^{t_2} dt\, \| e^{iHt} V \psi_0(t) \| = \int_{t_1}^{t_2} dt\, \| V \psi_0(t) \|. \tag{1.16}$$

Then using the bound

$$|\psi_0(x,t)| \leqslant \frac{c_1}{c_2 + |t|^{3/2}} \tag{1.17}$$

we end up with

$$\| [W(t_2) - W(t_1)]\, \psi_0(0) \| \leqslant \| V \| \int_{t_1}^{t_2} dt\, \frac{c_1}{c_2 + |t|^{3/2}} \leqslant c \| V \|. \tag{1.18}$$

Thus provided the potential has a finite norm

$$\| V \|^2 = \int dx\, V^2(x) < \infty \tag{1.19}$$

the Möller wave operator $\Omega^{(+)}$ defined in (1.13) exists. In fact even weaker conditions on V guarantee [1.3] the existence of $\Omega^{(+)}$. The potential has only to be locally square integrable and to decrease faster than the Coulomb potential at infinity.

The result achieved up to now is hardly surprising. We have only formulated and verified everyones expectation that the scattering solutions of the time dependent Schrödinger equation can be specified by certain initial conditions in the infinite past provided the potential is not too long range [see (1.19)]. In addition we have found a certain operator, $\Omega^{(+)}$, which maps the unperturbed initial state $|\psi_0\rangle$ into the complete state $|\Psi\rangle$.

The result (1.12) and (1.13) is not yet a practical one. The standard method of proceeding [1.4] is to reformulate it by using the relation:

$$\lim_{t \to -\infty} f(t) = \lim_{\varepsilon \to 0} \varepsilon \int_{-\infty}^{0} dt\, e^{\varepsilon t} f(t). \tag{1.20}$$

Exercise: Verify (1.20)

We then rewrite (1.13) together with (1.12) as

$$|\Psi(0)\rangle = \lim_{\varepsilon \to 0} \varepsilon \int_{-\infty}^{0} dt\, e^{\varepsilon t} e^{iHt} e^{-iH_0 t} |\psi_0(0)\rangle$$

$$= \lim_{\varepsilon \to 0} \varepsilon \int_{-\infty}^{0} dt\, e^{\varepsilon t} e^{iHt} \int dq\, e^{-iE_q t} |\psi_q^0\rangle f_0(q)$$

$$= \lim_{\varepsilon \to 0} \int dq\, \frac{i\varepsilon}{E_q + i\varepsilon - H} |\psi_q^0\rangle f_0(q). \tag{1.21}$$

In this manner we are led to an operator central to scattering theory: the resolvent operator to the Hamiltonian H

$$G(z) \equiv \frac{1}{z-H} \, . \tag{1.22}$$

Here z should obviously not be in the spectrum of H. Indeed in (1.21) $z = E_q + i\varepsilon$. We shall study properties of G in Sect. 1.3.

It is now tempting to apply G on a momentum eigenstate $|\psi_q^0\rangle$, which is of course not in Hilbert space. We define

$$|\Psi_q^{(+)}\rangle = \lim_{\varepsilon \to 0} \frac{i\varepsilon}{E_q + i\varepsilon - H} |\psi_q^0\rangle \tag{1.23}$$

and verify easily that $|\psi_q^{(+)}\rangle$ is a solution of the stationary Schrödinger equation

$$(H - E_q)|\Psi_q^{(+)}\rangle = 0 \, . \tag{1.24}$$

Since these states are not in the Hilbert space special care is needed in their use. Thus (1.23) is the operation by which stationary states to H_0, the momentum eigenstates, are mapped into specific eigenstates of H. The way $|\psi_q^{(+)}\rangle$ incorporates the features of the scattering process will be discussed in Sect. 1.4.

For a specific initial momentum the state $|\Psi_q^{(+)}\rangle$ contains all the information about the scattering process and we get the time dependent state for a general initial momentum distribution by superposition:

$$|\Psi(t)\rangle = e^{-iHt}|\Psi(0)\rangle = \int dq \, |\Psi_q^{(+)}\rangle e^{-iE_q t} f_0(q) \, . \tag{1.25}$$

1.2 The Cross Section

The main result of the last section, (1.25) together with (1.23) allows us to calculate the scattering state at all times. Specifically we can determine the transition amplitude at time t into a state

$$|\psi_{q_f}^0(t)\rangle = e^{-iH_0 t}|\psi_{q_f}^0(0)\rangle$$

of sharp momentum q_f:

$$A_{q_f}(t) \equiv \langle \psi_{q_f}^0(t)|\Psi(t)\rangle = \langle \psi_{q_f}^0(0)|e^{iH_0 t}e^{-iHt}|\Psi(0)\rangle. \tag{1.26}$$

Since the cross section is proportional to the transition rate, $(d/dt)|A|^2$, we shall also need

$$\dot{A}_{q_f}(t) = -i \langle \psi_{q_f}^0(0) | e^{iH_0 t} V e^{-iHt} | \Psi(0) \rangle .$$ (1.27)

Using (1.25) we find

$$A_{q_f}(t) = \int dq \, \exp[i(E_{q_f} - E_q)t] \langle \psi_{q_f}^0 | \Psi_q^{(+)} \rangle f_0(q)$$ (1.28)

and

$$\dot{A}_{q_f}(t) = -i \int dq \, \exp[i(E_{q_f} - E_q)t] \langle \psi_{q_f}^0 | V | \Psi_q^{(+)} \rangle f_0(q) .$$ (1.29)

What are the momentum components $\langle \psi_{q_f}^0 | \Psi_q^{(+)} \rangle$ of the stationary scattering state $| \Psi_q^{(+)} \rangle$, which is defined in (1.23)? If we switch off the potential the resolvent operator $G(z)$ turns into

$$G_0(z) \equiv \frac{1}{z - H_0}$$ (1.30)

and $| \Psi_q^{(+)} \rangle$ reduces to the momentum eigenstate $| \psi_q^0 \rangle$:

$$\lim_{\varepsilon \to 0} \frac{i\varepsilon}{E_q + i\varepsilon - H_0} | \psi_q^0 \rangle = | \psi_q^0 \rangle .$$ (1.31)

Then clearly we get

$$\langle \psi_{q_f}^0 | \psi_q^0 \rangle = \delta^3(q_f - q) ,$$ (1.32)

which inserted into (1.28) yields just the contribution to $A_{q_f}(t)$ from the unperturbed initial wave packet. How can we explicitly show that part in $| \Psi_q^{(+)} \rangle$? There is an obvious algebraic identity between the two resolvent operators $G_0(z)$ and $G(z)$:

$$G(z) = G_0(z) + G_0(z) V G(z) .$$ (1.33)

We use it in (1.23) to separate $| \Psi_q^{(+)} \rangle$ into a free and scattered part:

$$| \Psi_q^{(+)} \rangle = | \psi_q^0 \rangle + \lim_{\varepsilon \to 0} \frac{1}{E_q + i\varepsilon - H_0} V | \Psi_q^{(+)} \rangle .$$ (1.34)

Therefore we can express the momentum components of $| \Psi_q^{(+)} \rangle$ as

$$\langle \psi_{q_f}^0 | \Psi_q^{(+)} \rangle = \delta^3(q_f - q) + \lim_{\varepsilon \to 0} \frac{\langle \psi_{q_f}^0 | V | \Psi_q^{(+)} \rangle}{E_q + i\varepsilon - E_{q_f}} .$$ (1.35)

Here we encounter a central matrix element of scattering theory

$$T_{q_f q} \equiv \langle \psi^0_{q_f} | V | \Psi^{(+)}_q \rangle \qquad (1.36)$$

in terms of which we get

$$A_{q_f}(t) = f_0(q_f) + \lim_{\varepsilon \to 0} \int dq \, \exp[\mathrm{i}(E_{q_f} - E_q)t] \frac{T_{q_f q}}{E_q + \mathrm{i}\varepsilon - E_{q_f}} f_0(q) \qquad (1.37)$$

and

$$\dot{A}_{q_f}(t) = -\mathrm{i} \int dq \, \exp[\mathrm{i}(E_{q_f} - E_q)t] T_{q_f q} f_0(q) . \qquad (1.38)$$

Now we are prepared to calculate the transition rate at time t:

$$\frac{d}{dt}|A_{q_f}(t)|^2 = 2\,\mathrm{Re}\left\{ -\mathrm{i} \int dq \, \exp[\mathrm{i}(E_{q_f} - E_q)t] T_{q_f q} \right.$$

$$\times f_0(q) f_0^*(q_f) - \mathrm{i} \lim_{\varepsilon \to 0} \int dq \, \exp[\mathrm{i}(E_{q_f} - E_q)t]$$

$$\left. \times T_{q_f q} f_0(q) \int dq' \exp[-\mathrm{i}(E_{q_f} - E_{q'})t] \frac{T_{q_f q'}}{E_{q'} - \mathrm{i}\varepsilon - E_{q_f}} f_0^*(q') \right\} . \qquad (1.39)$$

We have to expect that it vanishes for large times t. For large times $|\Psi(t)\rangle$ describes the state when the particle has left the domain D and propagates again freely. Therefore the overlap $A_{q_f}(t) = \langle \psi^0_{q_f}(t) | \Psi(t) \rangle$ has to be time independent, since the two states belong to the same (free) Schrödinger equation. Indeed using the relation

$$\lim_{t \to \infty} \lim_{\varepsilon \to 0} \left(\frac{\mathrm{e}^{-\mathrm{i}xt}}{x + \mathrm{i}\varepsilon} \right) = -2\pi\mathrm{i}\,\delta(x) \qquad (1.40)$$

we get from (1.37)

$$\lim_{t \to \infty} A_{q_f}(t) = f_0(q_f) - 2\pi\mathrm{i} \int dq \, \delta(E_{q_f} - E_q) T_{q_f q} f_0(q)$$

$$= \int dq \, [\delta^3(q - q_f) - 2\pi\mathrm{i}\,\delta(E_{q_f} - E_q) T_{q_f q}] f_0(q)$$

$$\equiv \int dq \, S_{q_f q} f_0(q) . \qquad (1.41)$$

Clearly the quantity $S_{q_f q}$ is the probability amplitude for scattering from q to q_f and is called the S-matrix element. We shall say more about S in Sect. 1.5.

The probability $|A_{q_f}(t)|^2$ therefore approaches a time independent limit for $t \to \infty$ and its time derivative has to vanish. Mathematically this can also be seen directly from (1.39) using basic properties of Fourier transforms. Given this fact, how does a nonzero cross section arise? The cross section is the ratio of the transition rate to the incoming flux and we will now show that this ratio, as we go towards a stationary limit, will be nonzero. How do we approach the stationary situation in the initial state? The initial wave packet is given in (1.3). Normalized to 1 it describes the motion of one particle. This is reflected in the momentum distribution which sums up to 1:

$$\int dq \, |f_0(q)|^2 = 1 \,. \tag{1.42}$$

We introduced in (1.5a) as an example a Gaussian momentum distribution. In that example a decreasing value b will confine the momenta contained in the wave packet more and more to the neighbourhood of q_i. However because of the normalization condition (1.42) $f_0(q)$ cannot tend towards a δ-function. The normalization condition for a sequence of functions defining the δ-function is

$$\int f_\delta(q) \, dq = 1 \,, \tag{1.43}$$

which in the Gaussian form leads to

$$f_\delta(q) = \frac{1}{b^3} \frac{1}{(\pi)^{3/2}} e^{-(q-q_i)^2/b^2} \,. \tag{1.44}$$

Note the different powers in b occurring in (1.5a) and (1.44). We can write

$$f_0(q) = b^{3/2}(2\pi)^{3/4} f_\delta(q) \tag{1.45}$$

and the particle density in the Gaussian wave packet is expressed as

$$|\psi_0(x, t)|^2 = b^3(2\pi)^{3/2} |\int dq \, \psi_q^0(x) \, e^{-iE_q t} f_\delta(q)|^2 \,. \tag{1.46}$$

In the limit $b \to 0$ the property $f_\delta(q) \to \delta^3(q-q_i)$ reduces the integral to the plane wave state $\psi_{q_i}^0 \exp(-iE_{q_i}t)$ which has the constant particle density $(2\pi)^{-3}$. The factor $b^3(2\pi)^{3/2}$ therefore tells us by how much the probability to find the particle in a unit volume for a spreading wave packet is reduced in comparison to the constant probability of a plane wave state. As a consequence, the probability that the incoming particle hits the target of finite dimension and scatters into a final momentum state, described by $|A_{q_f}(t)|^2$, has to be expected to be reduced by the same factor. Indeed this is the case because of (1.45) and the quadratic dependence of $|A|^2$ on $f_0(q)$. The same is then true for the transition rate.

Now this rate, which decreases like b^3 for sharper and sharper energies, has to be divided by the incoming flux. The flux however, being of the form density × velocity, will also carry the factor b^3 in comparison to the constant flux j_0 belonging to a plane wave state. Indeed ($\overleftrightarrow{\nabla} \equiv \overrightarrow{\nabla} - \overleftarrow{\nabla}$)

$$j = \frac{1}{2im} (\psi_0^* \overleftrightarrow{\nabla} \psi_0) \tag{1.47}$$

and for $b \to 0$ we get

$$|j| \to b^3 (2\pi)^{3/2} j_0 \quad \text{with} \tag{1.48}$$

$$j_0 = \frac{|q_i|}{m} \frac{1}{(2\pi)^3} . \tag{1.49}$$

Therefore in the ratio between $(d/dt) |A_{q_f}(t)|^2$ and $|j|$ the factor $b^3 (2\pi)^{3/2}$ cancels and the stationary limit $b \to 0$ can be carried through. Thus instead of $(d/dt) |A|^2$ we regard $b^{-3} (d/dt) |A|^2$. Using (1.45) we derive from (1.39)

$$\lim_{b \to 0} \left(\frac{1}{b^3 (2\pi)^{3/2}} \frac{d}{dt} |A_{q_f}(t)|^2 \right)$$

$$= 2 \,\mathrm{Re} \left\{ -\mathrm{i} \delta^3 (q_f - q_i) T_{q_f q_i} - \mathrm{i} \lim_{\varepsilon \to 0} |T_{q_f q_i}|^2 \frac{1}{E_{q_i} - \mathrm{i}\varepsilon - E_{q_f}} \right\}$$

$$= 2 \,\mathrm{Im} \{ T_{q_f q_i} \delta^3 (q_f - q_i) + 2\pi \delta (E_{q_f} - E_{q_i}) |T_{q_f q_i}|^2 \} . \tag{1.50}$$

The first term results from the interference of the initial wave packet, the beam, with the scattered part of the wave function and is present only in the forward direction.

Let us now regard the scattering events which have a momentum different from the initial one. This is described by the second part, which moreover exhibits energy conservation, a property obviously expected in potential scattering. Now depending on the experimental set up we can calculate the number of events occurring per unit time. In potential scattering the most detailed observable is the number of particles scattered per unit time into a solid angle $d\hat{q}_f$ and into a small momentum interval Δq_f. Assuming constancy of $T_{q_f q_i}$ in these intervals that number is [up to the factor $(2\pi)^{3/2} b^3$, which will be cancelled by $|j|$]

$$dN = |T_{q_f q_i}|^2 d\hat{q}_f \int_{\Delta q_f} dq_f q_f^2 \, 2\pi \delta (E_{q_f} - E_{q_i}) = 2\pi m |q_i| |T_{q_f q_i}|^2 d\hat{q}_f . \tag{1.51}$$

Then the differential cross section

$$d\sigma \equiv \frac{dN (2\pi)^{3/2} b^3}{|j|} = \frac{dN}{j_0} \tag{1.52}$$

turns out to be

$$\frac{d\sigma}{d\hat{q}_f} = (2\pi)^4 m^2 |T_{q_f q_i}|^2 . \tag{1.53}$$

Let us summarize at this point. The most detailed observable in potential scattering, the differential cross section, is determined by the T-matrix element $T_{q_f q_i}$ defined in (1.36). Whereas in (1.36) q_f was not restricted, in the amplitude $T_{q_f q_i}$ entering into the cross section $E_{q_f} = E_{q_i}$. One talks in this case of the on-the-energy-shell amplitude or briefly of the on-shell amplitude, whereas in (1.36) the quantity unrestricted in q_f is called the half-on-the-energy-shell amplitude or briefly half-shell amplitude. This quantity will occur naturally in the Lippmann-Schwinger equation (1.127), the integral equation, with which one calculates the amplitude $T_{q_f q_i}$. We also want to emphasize the basic structure of the T-matrix element, which is quite general. It is built up from the scattering state $| \Psi_{q_i}^{(+)} \rangle$, the final state $| \psi_{q_f}^0 \rangle$ and an interaction V. The scattering state arises from a certain choice of initial condition and the final state is an eigenstate of the Hamiltonian $(H - V)$, where V is the interaction. Finally the scattering state is defined through (1.23).

Let us now return to (1.50). The probabilities $|A_{q_f}|^2$ have to add up to unity if summed over all final momenta. Therefore the time rate of that sum has to vanish. Consequently we can conclude from (1.50)

$$\int dq_f \, 2\pi \, \delta(E_{q_i} - E_{q_f}) |T_{q_f q_i}|^2 = -2 \int dq_f \delta^3(q_f - q_i) \, \mathrm{Im} \{ T_{q_i q_i} \}$$
$$= -2 \, \mathrm{Im} \{ T_{q_f q_i} \} \,. \tag{1.54}$$

On the lhs we recognize essentially the total cross section

$$\sigma_{\mathrm{tot}} \equiv \int d\hat{q}_f \frac{d\sigma}{d\hat{q}_f} = (2\pi)^4 m^2 \int d\hat{q}_f |T_{q_f q_i}|^2 \tag{1.55}$$

and we find

$$\mathrm{Im} \{ T_{q_i q_i} \} = -\frac{1}{2} \frac{|q_i|}{m(2\pi)^3} \sigma_{\mathrm{tot}} \,. \tag{1.56}$$

This relation is known as the optical theorem and is of great importance in evaluating dispersion relations.

1.3 Resolvent Operators and Green's Functions

This section serves to exhibit some properties of resolvent operators or of Green's functions as their coordinate representations are often called. They play an essential role in the formulation of scattering theories. From the Hamilton operators H_0 and H, which we have encountered up to now, one can form the two resolvent operators $G_0(z)$ and $G(z)$ as given in (1.30) and (1.22). They way they act can be exhibited in the following manner. Let us denote the eigenstates and eigenvalues of H_0 by $|q\rangle$ and E_q, respectively. We have already introduced the coordinate representation of the states $|q\rangle$, namely

$$\langle x | q \rangle \equiv \psi_q^0(x) = \frac{1}{(2\pi)^{3/2}} e^{iqx} . \tag{1.57}$$

Therefore they are normalised as

$$\langle q' | q \rangle = \delta^3(q - q') \tag{1.58}$$

and span the total space

$$\int dq |q\rangle\langle q| = 1 . \tag{1.59}$$

Then the spectral representation of $G_0(z)$ is

$$G_0(z) = \int dq |q\rangle \frac{1}{z - E_q} \langle q| . \tag{1.60}$$

We see that $G_0(z)$ is defined for all values of z which do not lie in the spectrum of H_0, which is $0 \leqslant E_q < \infty$. However we can approach with z towards the spectrum of H_0 and due to

$$\lim_{\varepsilon \to 0} \frac{1}{E \pm i\varepsilon - E_q} = \frac{P}{E - E_q} \mp i\pi \delta(E - E_q) \tag{1.61}$$

the limits

$$G_0^{(\pm)}(E) \equiv \lim_{\varepsilon \to 0} G_0(E \pm i\varepsilon) \tag{1.62}$$

exist. Moreover they are different:

$$G_0^{(+)}(E) - G_0^{(-)}(E) = -2i\pi \int dq |q\rangle \delta(E - E_q)\langle q| . \tag{1.63}$$

Therefore $G_0(z)$ is defined on the complex z-plane, which is cut along the positive real axis.

Now we can regard $G(z)$. The Hamilton operator H has in general two types of eigenstates, scattering states and bound states. We encountered in Sect. (1.1) the states

$$|q\rangle^{(+)} \equiv \Omega^{(+)} |q\rangle \tag{1.64}$$

as defined in (1.23). They are eigenstates of H to the eigenvalues $0 \leqslant E_q < \infty$. Their normalisation is the same as for the states $|q\rangle$:

$$^{(+)}\langle q' | q \rangle^{(+)} = \delta^3(q - q') . \tag{1.65}$$

Therefore they are not elements of Hilbert space and the phrase "generalized eigenstates" should be used instead. In fact we shall encounter for instance at the end of Sect. 3.2 nonzero surface terms which stem just from the oscillatory behaviour at infinity of this type of "generalized eigenstates". Nevertheless we shall use throughout this book the phrase eigenstate. The property (1.165) can be seen in the following manner. Let us consider the norm of the scattering state $|\Psi(t)\rangle$. It is independent of time and because of the initial condition (1.9) we can calculate it at $t \to -\infty$.

$$
\begin{aligned}
\langle \Psi(t)|\Psi(t)\rangle &= \lim_{\tau \to -\infty} \langle \Psi(\tau)|\Psi(\tau)\rangle \\
&= \lim_{\tau \to -\infty} \langle \psi_0(\tau)|\psi_0(\tau)\rangle = \langle \psi_0(t)|\psi_0(t)\rangle .
\end{aligned}
\tag{1.66}
$$

The equalities of the norms at $\tau \to -\infty$ follow from (1.9) through an elementary estimate using Schwarz inequality. In the last equality we again applied the time-independence now of the norm of ψ_0. Thus

$$
\langle \Psi(t)|\Psi(t)\rangle = \int dq \, |f_0(q)|^2 .
\tag{1.67}
$$

On the other hand according to (1.25) we have

$$
\langle \Psi(t)|\Psi(t)\rangle = \int dq \int dq'^{(+)}\langle q|q'\rangle^{(+)} e^{i(E_q - E_{q'})t} f_0^*(q) f_0(q') .
\tag{1.68}
$$

The two forms (1.67, 68) are only compatible if (1.65) holds.

In order to familiarize ourselves with algebraic manipulations we present another way to verify (1.65). Let us use the resolvent identity between G and G_0 now however in the form

$$
G(z) = G_0(z) + G(z) V G_0(z) .
\tag{1.69}
$$

The we can evaluate (1.23) with the result

$$
|q\rangle^{(+)} = |q\rangle + \lim_{\varepsilon \to 0} G(E_q + i\varepsilon) V |q\rangle .
\tag{1.70}
$$

In the dual space this reads

$$
{}^{(+)}\langle q| = \langle q| + \lim_{\varepsilon \to 0} \langle q| V G(E_q - i\varepsilon) .
\tag{1.71}
$$

Then we can proceed:

$$
\begin{aligned}
{}^{(+)}\langle q'|q\rangle^{(+)} &= \langle q'|q\rangle^{(+)} + \lim_{\varepsilon \to 0} \langle q'|V G(E_{q'} - i\varepsilon)|q^{(+)}\rangle \\
&= \delta^3(q - q') + \lim_{\varepsilon \to 0} \frac{1}{E_q + i\varepsilon - E_{q'}} \langle q'|V|q\rangle^{(+)} \\
&\quad + \lim_{\varepsilon \to 0} \langle q'|V|q\rangle^{(+)} \frac{1}{E_{q'} - i\varepsilon - E_q} = \delta^3(q - q') .
\end{aligned}
\tag{1.72}
$$

In the second equality we used (1.34) and the fact that $|q\rangle^{(+)}$ is an eigenstate of H.

The potential V, if it is strong enough, can also support a second type of states, square integrable ones, which describe bound states $|b\rangle$ at discrete negative energies $E_b < 0$. These two types of states span [1.5] again the total space available to the particle

$$\sum_b |b\rangle\langle b| + \int dq\, |q\rangle^{(+)}\,^{(+)}\langle q| = 1 \,. \tag{1.73}$$

Therefore the spectral representation of $G(z)$ is

$$G(z) = \sum_b |b\rangle \frac{1}{z - E_b} \langle b| + \int dq\, |q\rangle^{(+)} \frac{1}{z - E_q}\,^{(+)}\langle q| \,. \tag{1.74}$$

Comparing with $G_0(z)$ in (1.60) we see that $G(z)$ is also defined on the complex z-plane cut along the real axis, but it can also have additional poles at $z = E_b < 0$. These bound state poles of $G(z)$ will play an important dynamical role in approximating two body transition operators as we shall see in Chap. 3.

The property of the resolvent operators of acting as propagators for the motion of the particle is seen in the coordinate representation. In that representation the resolvent operators are called Green's functions. Let us regard the free Green's function

$$G_0(x, x', z) \equiv \langle x|G_0(z)|x'\rangle \,. \tag{1.75}$$

Inserting the spectral representation (1.60) we find

$$G_0(x, x', z) = \frac{1}{(2\pi)^3} \int dq\, \frac{\exp[iq(x-x')]}{z - E_q} \,. \tag{1.76}$$

It is a simple exercise to calculate that integral and one gets

$$G_0(x, x', z) = -\frac{m}{2\pi} \frac{\exp(i\sqrt{2mz}\,|x-x'|)}{|x-x'|} \,. \tag{1.77}$$

From that explicit expression we indeed see that the limits $z = E \pm i\varepsilon$, $\varepsilon \to 0$ exist and are different. Specifically we find that the cut is caused by a square root branch point. For positive energies, G_0 describes the free propagation from x' to x, as will be apparent in the context of (1.81) in the next section. For this reason G_0 is often called the free propagator. One verifies easily that $G_0(x, x', z)$ obeys the equation

$$\left(-\frac{\nabla_x^2}{2m} - z\right) G_0(x, x', z) = -\delta^3(x-x') \,, \tag{1.78}$$

which is of course nothing else than the coordinate representation of

$$(H_0 - z)G_0(z) = -1.$$ (1.79)

The full Green's function $G(x, x', z)$ describes the propagation of the particle under the influence of V. Since $V(x)$ destroys translation invariance, it will depend on both x and x' rather than the difference $x - x'$. In a partial wave decomposition useful representations of G are available, which do not involve the integral in the spectral representation (1.74) (see for instance [1.5]).

1.4 Asymptotic Behaviour of the Scattering Wave Function

Equipped with the coordinate representation of the free Green's function $G_0(x, x', z)$ we can follow the particle motion in space and time. The stationary scattering wave function

$$\Psi_q^{(+)}(x) \equiv \langle x | q \rangle^{(+)}$$ (1.80)

obeys the integral equation (1.34), which is called the Lippmann-Schwinger equation [1.6]. In coordinate representation it is

$$\Psi_q^{(+)}(x) = \psi_q^0(x) - \frac{m}{2\pi} \int dx' \, \frac{\exp(i\sqrt{2mE}\,|x - x'|)}{|x - x'|} \, V(x') \, \Psi_q^{(+)}(x').$$ (1.81)

The integrand can be thought of as describing the action of V at the position x' followed by a free propagation to the position x. That propagation is fed by the wave function $\Psi_q^{(+)}(x')$ itself. The content of (1.81) is more easily grasped by regarding the related perturbation series, obtained by replacing $\Psi_q^{(+)}(x')$ by the right hand side of (1.81) again and again. Then the general term in that series describes the situation that the free wave ψ_q^0 hits V at x_1, propagates freely towards x_2 where it hits V again and propagates towards x_3 and so on until it reaches the point x.

In this section we shall be mainly concerned with the behaviour of $\Psi_q^{(+)}(x)$ for large $|x|$ values. This behaviour is contained in (1.81). Whereas the Schrödinger equation (1.24) has to be supplemented by boundary conditions to single out specific solutions, this specification is already incorporated in the integral equation (1.81). Obviously this has to be the case if (1.81) specifies the solution uniquely. Assume there would be a second solution. Then the difference of the two solutions would obey the homogeneous equation related to (1.81). We shall show below [see (1.90)] that the integral in (1.81) behaves asymptotically like $\exp(i\sqrt{2mE}\,x)/x$ for $x \to \infty$, which leads to a nonzero flux through a sphere of radius x. In (1.81) the inhomogeneous term is the source of this flux. In the related homogeneous equation this source for the flux is absent, and since the kernel is nonsingular there is no other source possible. Since, moreover, the solutions of the stationary Schrödinger equation are flux conserving, there would be a contradiction, and we conclude that the solution must be unique. Only in the case of a bound state at $E < 0$ the homogeneous equation related to (1.81) has a nonzero solution. Then $\exp(i\sqrt{2mEx})/x$ is exponentially decreasing and the asymptotic flux through a sphere is zero of course.

The uniqueness of the solution for (1.81) is not surprising. First we note by operating with $[(-\nabla^2/2m) - E]$ on both sides of (1.81) and using (1.78) that every solution of (1.81) is a solution of the stationary Schrödinger equation to the energy $E > 0$. However, on physical grounds there exists only one scattering state, $\Psi_q^{(+)}(x)$, linked to an initial momentum q. We emphasize this point here since we shall see in Chap. 3, that for more than two particles there are in general different scattering states at the same energy and the Lippmann-Schwinger equation does not specify all boundary conditions and consequently allows in general several solutions. The reason is that the corresponding kernel of the Lippmann Schwinger equation does not have the property to produce only outgoing waves.

The time-dependent scattering wave function is given through

$$\Psi(x, t) = \int dq\, \Psi_q^{(+)}(x, t)\, e^{-iE_q t} f_0(q) \equiv \psi_0(x, t) + \psi_{\text{scatt}}(x, t) . \tag{1.82}$$

Let us first regard the unperturbed initial wave packet $\psi_0(x, t)$ resulting from the driving term in (1.81):

$$\psi_0(x, t) = \int dq\, \frac{1}{(2\pi)^{3/2}} \exp\left[i(qx - E_q t)\right] f_0(q) . \tag{1.83}$$

For large times $|t|$ and fixed x $\psi_0(x, t)$ decreases like $|1/t|^{3/2}$ as we have already seen in Sect. 1.1. The reason is simply the oscillatory behaviour of the integrand. The decrease will be slower if x and t are correlated such that the phase varies as little as possible. This idea of stationary phase approximation [1.7] is included if we proceed in the following manner. We assume that $f_0(q)$ depends on $(q - q_i)$ and is peaked at q_i. Then we expand the exponent in (1.83) around q_i, put $\varrho = q - q_i$ and get

$$\psi_0(x, t) = \psi_{q_i}^0(x, t) \int d\varrho\, \exp\left[i\varrho(x - (q_i/m)\, t)\right]\, e^{-i(\varrho^2/2m)t} f_0(\varrho) . \tag{1.84}$$

The exponent in the second exponential is just of the order ξ defined in (1.5c), where b is now a measure of the width of $f_0(\varrho)$. We shall again neglect that spreading term and find immediately, introducing the Fourier transform $\tilde{f}_0$,

$$\psi_0(x, t) \approx \psi_{q_i}^0(x, t)\tilde{f}_0\left(x - \frac{q_i}{m}\, t\right) . \tag{1.85}$$

In this approximation the wave packet does not change its form and its center travels with the group velocity $v = q_i/m$. Within the domain spanned by $\tilde{f}_0$ the time dependent plane wave state $w^0_{q_i}(x, t)$ propagates with the phase velocity $v_{ph} = v/2$. Furthermore the beam particle density at position x and at time t is

$$\varrho(x, t) = \frac{1}{(2\pi)^3}\left|\tilde{f}_0\left(x - \frac{q_i}{m}t\right)\right|^2. \tag{1.86}$$

Let us choose the z-axis along the q_i direction and put the origin at the center of V. Also we assume that the width of $\tilde{f}_0$ is large with respect to the size of D so that the variation of $\tilde{f}_0(x)$ over D is negligible. Then the incident flux per unit area orthogonal to the beam direction is

$$F_0 = \int\limits_{-\infty}^{\infty} dz \frac{1}{(2\pi)^3}|\tilde{f}_0(0, 0, z)|^2. \tag{1.87}$$

We arrive at the same conclusion of course in the standard manner using (1.85) in calculating j_z according to (1.47). Neglecting the transient contributions induced by the change of $\tilde{f}_0$ at its borders we get

$$j_z = \frac{q_i}{m} \frac{1}{(2\pi)^3}\left|\tilde{f}_0\left(0, 0, z - \frac{q_i}{m}t\right)\right|^2, \tag{1.88}$$

which integrated over all times yields (1.87).

Let us now regard the scattered part defined in (1.82). Clearly to find the leading contribution for large times t we have again to correlate x and t. Thus the asymptotic behaviour of the stationary scattering wave function $\Psi^{(+)}_q(x)$ for $|x| \to \infty$ is needed. Since we assumed that $V(x)$ is of finite range the value of x in (1.81) can be assumed to be large with respect to x' and we get

$$|x - x'| = x - x' \, \hat{x}\hat{x}' + O\left(\frac{1}{|x|}\right). \tag{1.89}$$

Therefore one finds

$$\Psi^{(+)}_q(x) \xrightarrow[|x| \to \infty]{} \frac{1}{(2\pi)^{3/2}}\left[e^{iqx} + \frac{e^{iqx}}{x}f(q_f, q)\right] \tag{1.90}$$

with

$$f(q_f, q) = -m\sqrt{2\pi} \int dx \, e^{-iq_f x} V(x) \Psi^{(+)}_q(x). \tag{1.91}$$

We introduced the final momentum q_f as pointing into the direction $\hat{x}$ of observation and having the magnitude q_i:

$$q_f \equiv q_i \hat{x} \, . \tag{1.92}$$

Now we can discuss the asymptotic form of $\psi_{scatt}(x, t)$ for both $|x|$ and $|t|$ being large:

$$\psi_{scatt}(x, t) \rightarrow \frac{1}{(2\pi)^{3/2}} \int dq \, \frac{e^{iqx}}{x} f(q_f, q) \, e^{-iE_q t} f_0(q) \, . \tag{1.93}$$

First of all for $t \rightarrow -\infty$ small phases are not possible and the right hand side vanishes. Of course this has to be the case since $|\Psi(t)\rangle$ has to tend towards $|\psi_0(t)\rangle$ for $t \rightarrow -\infty$. We now evaluate (1.93) in a similar manner to (1.83). We again expand the exponents around q_i and get in first order

$$e^{iqx} e^{-iE_q t} = e^{iq_i x} e^{-iE_{q_i} t} \exp\left[i(q - q_i)\left(\hat{q}_i x - \frac{q_i}{m} t\right)\right] \, . \tag{1.94}$$

We shall neglect the variations of $f(q_f q)$ over the width of $f_0(q)$ and thus the explicit appearance of a time shift with respect to the potential free motion. Then (1.93) turns into

$$\psi_{scatt}(x, t) \simeq \frac{1}{(2\pi)^{3/2}} \frac{e^{iq_i x}}{x} f(q_f q_i) \, e^{-iE_{q_i} t} \tilde{f}_0\left(\hat{q}_i x - \frac{q_i}{m} t\right) \, . \tag{1.95}$$

The wave packet appears now as a spherical shell whose average radius increases according to the classical particle velocity $v = |q_i|/m$. Inside that shell we encounter a spherical wave $x^{-1} \exp(iq_i x) \exp(-iE_{q_i} t)$ which is moreover modulated in the various directions $\hat{q}_f$ by the amplitude $f(q_f, q_i)$. The form (1.95) leads immediately to the total flux per beam particle going through a unit area in the direction $\hat{x}$:

$$F = \hat{x} \, \frac{1}{x^2} \, \frac{q_i}{m} \, \frac{1}{(2\pi)^3} |f(q_f q_i)|^2 \int_{-\infty}^{+\infty} dt \, f_0^2\left(0, 0, z - \frac{q_i}{m} t\right)$$

$$= \hat{x} \, \frac{1}{x^2} |f|^2 F_0 \, . \tag{1.96}$$

In the last equality we encounter the initial flux per unit area of (1.87). Consequently the number of scattering events per beam particle passing through the surface element $df = \hat{x} \, x^2 d\hat{x} = \hat{x} x^2 d\hat{q}_f$ is

$$dF = F \, df = |f(q_f q_i)|^2 F_0 d\hat{q}_f \tag{1.97}$$

and the differential cross section

$$d\sigma \equiv \frac{dF}{F_0} = |f(q_f, q_i)|^2 \, d\hat{q}_f \, . \tag{1.98}$$

Recalling the definitions (1.91) and (1.36) of the scattering amplitude f and the T-matrix element, respectively, we see that the two expressions (1.98) and (1.53) for the differential cross section agree.

As a short cut for arriving at (1.98) one uses directly the asymptotic form (1.90) of the stationary scattering wavefunction $\Psi_q^{(+)}(x)$. The plane wave stands for the initial beam and the spherically outgoing wave (it is outgoing due to the time factor $\exp(-iE_q t)$ for the scattered part. The radially scattered flux density is obviously $q_i/m|f|^2$ and the flux density of the plane wave q_i/m. The ratio yields immediately (1.98).

Let us end this section with a remark on flux conservation calculated via this short cut. The conservation of probability which led to (1.54) can be rewritten with the aid of (1.91) and (1.36) in terms of the on-shell quantity f as

$$q_i \int_{-1}^{1} d\cos\vartheta |f(\cos\vartheta)|^2 - 2\operatorname{Im}\{f(1)\} = 0 . \tag{1.99}$$

Note that $f(q_f, q_i)$ resulting from a rotationally invariant potential V, must also be rotationally invariant. It can therefore depend only on $\cos\vartheta \equiv \hat{q}_f \hat{q}_i$.

Now we want to show that (1.99) can also be obtained if we consider the flux through a sphere of radius r outside D and require it to be zero. Since we shall use the asymptotic form (1.90) we choose a radius $r \to \infty$.

Up to a normalization Ψ and $\partial\Psi/\partial r$ have the asymptotic form

$$\Psi \to \exp(iq_i r \cos\vartheta) + \frac{e^{iq_i r}}{r} f(\cos\vartheta) + O\left(\frac{1}{r^2}\right) \tag{1.100}$$

and

$$\frac{\partial\Psi}{\partial r} \to iq_i \cos\vartheta \exp(iq_i r \cos\vartheta) + iq_i \frac{e^{iq_i r}}{r} f(\cos\vartheta) - \frac{e^{iq_i r}}{r^2} f(\cos\vartheta) . \tag{1.101}$$

Since we have to evaluate

$$F \equiv \lim_{r\to\infty} \int d\Omega \, j_r r^2$$

higher order terms in r^{-1} will not contribute. One finds

$$j_r = \frac{q_i}{m} \cos\vartheta + \frac{q_i}{2m} (1 + \cos\vartheta) \frac{\exp[iq_i r(1 - \cos\vartheta)]}{r} f(\cos\vartheta)$$

$$+ \frac{q_i}{2m} (1 + \cos\vartheta) \frac{\exp[-iq_i r(1 - \cos\vartheta)]}{r} f^*(\cos\vartheta)$$

$$+ \frac{q_i}{m} |f(\cos\vartheta)|^2 \frac{1}{r^2} - \frac{1}{2im} \frac{1}{r^2} \{\exp[iq_i r(1 - \cos\vartheta)] f(\cos\vartheta)$$

$$- \exp[-iq_i r(1 - \cos \vartheta)] f^*(\cos \vartheta)\}$$

$$+ \frac{q_i}{2m} \cos \vartheta \left[O\left(\frac{1}{r^2}\right) \exp(iq_i r \cos \vartheta) + \text{conj. compl.} \right] + O\left(\frac{1}{r^3}\right).$$

$$(1.102)$$

The second term on the rhs will lead to an integral $(t \equiv \cos \vartheta)$

$$I(r) \equiv \frac{1}{r} \int_{-1}^{1} dt (1 + t) \exp[iq_i r(1 - t)] f(t).$$

$$(1.103)$$

For $r \to \infty$ the leading contribution results from the upper limit $t = 1$ [if $f(t)$ is well behaved]. One easily finds or $r \to \infty$

$$I(r) = \frac{-2}{iq_i r^2} f(1) + O\left(\frac{1}{r^3}\right).$$

$$(1.104)$$

Therefore this second term and also the third term will contribute to the asymptotic flux. They result from the interference between the plane wave and the scattered part. Clearly the scattered part, the fourth term, will also contribute, whereas the following ones when integrated over t are of the order $O(r^{-3})$ and will not contribute. Finally the first term, the plane wave part, is flux conserving by itself. Thus putting $F = 0$ leads indeed to (1.99). We shall recover that relation again in the next section as the unitarity relation of the S-matrix.

1.5 The S-, T-, and K-Matrices

In a scattering process the initial and final state is accessible to a measurement. Thus the probability amplitude for a transition from q_i to q_f in the case of potential scattering is the central quantity of interest. It was determined in Sect. 1.2, Eq. (1.41), as

$$S_{q_f q_i} = \delta^3(q_f - q_i) - 2\pi i \, \delta(E_{q_f} - E_{q_i}) \, T_{q_f q_i}.$$

$$(1.105)$$

Let us rederive this expression by introducing a second type of scattering state, an auxiliary mathematical one. Instead of regarding a solution of the time dependent Schrödinger equation which develops out of a free state in the infinite past, as we did in Sect. 1.1, one can introduce a solution which develops backwards in time out of a specific free state in the infinite future. In exactly the same manner as in Sect. 1.1 we can find the corresponding Möller wave operator, now called $\Omega^{(-)}$, which maps the free state into the new type of solution of the full Schrödinger equation. To distinguish the two types we shall denote them by $\Psi^{(\pm)}(t)$ and they are given by

$$|\Psi^{(\pm)}(t)\rangle = \Omega^{(\pm)}|\psi_0(t)\rangle . \tag{1.106}$$

Clearly the new Möller wave operator is

$$\Omega^{(-)} = \lim_{\tau \to +\infty} e^{iH\tau}e^{-iH_0\tau} \tag{1.107}$$

and its existence on the whole Hilbert space can be proved exactly as in Sect. 1.1. By their very definition the two states have the property

$$\|\Psi^{(\pm)}(t) - \psi_0(t)\| \to 0 \quad \text{for} \quad t \to \mp\infty . \tag{1.108}$$

As $|q\rangle^{(+)}$ results from $|q\rangle$ through $\Omega^{(+)}$, a second type of eigenstate of the time independent Schrödinger equation can be created through $\Omega^{(-)}$:

$$|q\rangle^{(-)} = \lim_{\varepsilon \to 0} \frac{-i\varepsilon}{E_q - i\varepsilon - H}|q\rangle . \tag{1.109}$$

It will obey the Lippmann-Schwinger equation

$$|q\rangle^{(-)} = |q\rangle + \lim_{\varepsilon \to 0} \frac{1}{E_q - i\varepsilon - H_0} V|q\rangle^{(-)}, \tag{1.110}$$

which has the formal solution

$$|q\rangle^{(-)} = |q\rangle + \lim_{\varepsilon \to 0} \frac{1}{E_q - i\varepsilon - H} V|q\rangle . \tag{1.111}$$

Now we are prepared to ask again for the probability amplitude for a transition from an initial free state, $|\psi_{0_i}(t)\rangle$, to a final state, $|\psi_{0_f}(t)\rangle$, where i and f denote specific wave packets. It is given by

$$A_{fi} = \lim_{t \to +\infty} \langle \psi_{0_f}(t)|\Psi_i^{(+)}(t)\rangle . \tag{1.112}$$

Now however we can use (1.108) and get

$$A_{fi} = \lim_{t \to +\infty} \langle \Psi_f^{(-)}(t)|\Psi_i^{(+)}(t)\rangle = \langle \psi_f^{(-)}(0)|\Psi_i^{(+)}(0)\rangle . \tag{1.113}$$

Again we have exploited the time-independence of the matrix elements in (1.113). In explicit notation this reads

$$A_{\mathrm{fi}} = \int dq \int dq' f_{0_{\mathrm{f}}}^*(q) f_{0_{\mathrm{i}}}(q')\, {}^{(-)}\langle q|q'\rangle^{(+)}, \tag{1.114}$$

which reveals the probability amplitude between sharp momentum states

$$S_{qq'} \equiv {}^{(-)}\langle q|q'\rangle^{(+)}. \tag{1.115}$$

We can establish the equality with (1.105) in the following straightforward manner:

$$\begin{aligned}
S_{qq'} &= \langle q|q'\rangle^{(+)} + \lim_{\varepsilon \to 0}\left\langle q\left|V\frac{1}{E_q + i\varepsilon - H}\right|q'\right\rangle^{(+)} \\
&= \langle q|q'\rangle + \lim_{\varepsilon \to 0}\frac{1}{E_{q'} + i\varepsilon - E_q}\langle q|V|q'\rangle^{(+)} \\
&\quad + \lim_{\varepsilon \to 0}\frac{1}{E_q + i\varepsilon - E_{q'}}\langle q|V|q'\rangle^{(+)} \\
&= \delta^3(q - q') - 2i\pi\,\delta(E_q - E_{q'})\,T_{qq'}. \tag{1.116}
\end{aligned}$$

The form (1.115) also tells us immediately that the S-matrix must be unitary. Why? The completeness relation (1.73) can be equally well written with the aid of the $|q\rangle^{(-)}$ states:

$$\sum |b\rangle\langle b| + \int dq\, |q\rangle^{(-)}\, {}^{(-)}\langle q| = 1 . \tag{1.117}$$

Also the $|q\rangle^{(-)}$ states are orthonormalized as the $|q\rangle^{(+)}$ states and of course are orthogonal to the bound states $|b\rangle$. Now using (1.73) and (1.117) we can express $|q\rangle^{(+)}$ in terms of $|q\rangle^{(-)}$ states or $|q\rangle^{(-)}$ in terms of $|q\rangle^{(+)}$ states. The expansion coefficients are just $S_{q'q}$ or $S_{qq'}^*$ respectively, and the unitarity relations

$$\int dq\, S_{qq'}^* S_{qq''} = \delta^3(q - q') = \int dq\, S_{q'q} S_{q''q}^* \tag{1.118}$$

results from the orthonormality of $|q\rangle^{(+)}$.

Let us now work out the connection with the flux conserving relation (1.99), which appears in the form (1.54) when expressed in terms of the T-matrix elements $T_{qq'}$. We insert (1.105) into (1.118) and find easily

$$\mathrm{i}(T_{q''q'}^* - T_{q'q''}) + 2\pi \int dq\,\delta(E_q - E_{q'})\, T_{qq'}^* T_{qq''} = 0 \tag{1.119}$$

and

$$\mathrm{i}(T_{q'q''}^* - T_{q''q'}) + 2\pi \int dq\,\delta(E_q - E_{q'})\, T_{q''q} T_{q'q}^* = 0 . \tag{1.120}$$

Note that all momentum states have the same energy: $E_q = E_{q'} = E_{q''}$.

The appearance of two similar looking equations forces us to work out the possible relation between $T_{qq'}$ and $T_{q'q}$. Intuitively we expect a relation linked to time reversal. Let us assume that the potential is invariant under time reversal and let us denote the antiunitary operator for the time reversal operation [1.8] by $\mathcal{T}$. Then $\mathcal{T}$ commutes with V,

$$[\mathcal{T}, V] = 0 \tag{1.121}$$

and it will map the state $|q\rangle$ into the time reversed state $|-q\rangle$:

$$\mathcal{T}|q\rangle = |-q\rangle. \tag{1.122}$$

What are the consequences for $T_{qq'}$? It is explicitly given by

$$T_{qq'} = \langle q|V|q'\rangle + \lim_{\varepsilon \to 0}\langle q|V\frac{1}{E_{q'}+i\varepsilon-H}V|q'\rangle. \tag{1.123}$$

Because of the antiunitary nature of $\mathcal{T}$ we can reexpress the rhs as

$$T_{qq'} = \langle \mathcal{T}q|\mathcal{T}V|q'\rangle^* + \lim_{\varepsilon \to 0}\langle \mathcal{T}q|\mathcal{T}V\frac{1}{E_{q'}+i\varepsilon-H}V|q'\rangle^*. \tag{1.124}$$

Then using (1.121) and (1.122) and recalling the on-shell nature $E_q = E_{q'}$ we get

$$T_{qq'} = \langle -q|V|-q'\rangle^* + \lim_{\varepsilon \to 0}\langle -q|V\frac{1}{E_{q'}-i\varepsilon-H}V|-q'\rangle^*$$

$$= \langle -q'|V|-q\rangle + \lim_{\varepsilon \to 0}\langle q'|V\frac{1}{E_q+i\varepsilon-H}V|-q\rangle = T_{-q', -q}. \tag{1.125}$$

Thus, as possibly expected, the T-matrix elements for time reversed processes are equal. Later we shall see that this property in a partial wave basis leads to the symmetry of the T- or S-matrix.

Due to the equality (1.125) and since q'' and q' are arbitrary momenta the second relation (1.120) follows from the first if time reversal invariance is valid. For $q' = q''$ the physical content of the unitarity relation (1.119) is obvious. Comparing with (1.54) or (1.99) it is just flux conservation.

After all this formal development let us now be concerned with what is finally needed in a practical application. The cross section is determined through the T-matrix (1.36). It is natural to ask for an integral equation directly for that quantity. The T-matrix element contains $|q\rangle^{(+)}$ which obeys the Lippmann-Schwinger equation (1.34). Therefore $T_{qq'}$ can be expressed as

$$T_{qq'} = \langle q|V|q'\rangle + \lim_{\varepsilon \to 0}\langle q|VG_0(E_{q'}+i\varepsilon)V|q'\rangle^{(+)}. \tag{1.126}$$

This is indeed an integral equation for $T_{qq'}$ since if we recall the spectral representation (1.60) of G_0 in terms of momentum eigenstates, (1.126) reads explicitely:

$$T_{qq'} = \langle q|V|q'\rangle + \lim_{\varepsilon \to 0} \int dq'' \langle q|V|q''\rangle \frac{T_{q''q'}}{E_{q'} + i\varepsilon - E_{q''}} . \qquad (1.127)$$

This equation is known as the Lippmann-Schwinger equation for the T-matrix. It connects the half-shell T-matrix elements with each other. Once it is solved the on-shell matrix element is also known, and one can calculate the cross section.

For formal developments it is advisable to introduce the T-operator and to generalize at the same time (1.127) replacing the energy $E_{q'} + i\varepsilon$ by an arbitrary complex value z. Then (1.127) is just the momentum representation of the operator equation

$$T(z) = V + VG_0(z)T(z) . \qquad (1.128)$$

In other words

$$T_{qq'}(z) \equiv \langle q|T(z)|q'\rangle . \qquad (1.129)$$

Equation (1.127) for complex z or the operator version (1.128) defines the off-shell T-matrix. This extension will be needed in the description of systems with more than two particles, as we shall see in Chaps. 3 and 4. Once the new independent energy variable z has been introduced, one can investigate the analytic properties of $T(z)$. The outcome of the following brief study will have important consequences for the approximate treatment of the off-shell T-operator.

Let us play with (1.128) by iterating it. The result is a series with increasing powers of V:

$$T(z) = V + VG_0(z)V + VG_0(z)VG_0(z)V + \dots . \qquad (1.130)$$

On the rhs we recognize the series

$$G_0 + G_0VG_0 + G_0VG_0VG_0 + \dots \qquad (1.131)$$

gained in iterating the resolvent identity (1.33) for $G(z)$. Therefore we can sum up the infinite series in (1.130) into

$$T(z) = V + VG(z)V . \qquad (1.132)$$

This is the formal solution of the Lippmann-Schwinger equation (1.128). We also arrive at the same result of course, in a rigorous and purely algebraic manner, using the resolvent identity (1.69) in the form

$$[1 + VG(z)][1 - VG_0(z)] = 1 . \qquad (1.133)$$

Then

$$[1 - VG_0(z)] T(z) = V \tag{1.134}$$

turns indeed into

$$T(z) = [1 + VG(z)] V. \tag{1.135}$$

The expression (1.132) exhibits important analytic properties of $T(z)$. As we saw in Sect. 1.3 $G(z)$ is defined on the complex z-plane cut along $0 \leqslant z < \infty$ due to the continuous spectrum of H and has in general poles at the discrete bound state energies $E_b < 0$. These properties obviously carry over directly to $T(z)$. Moreover the residues of $T(z)$ factorise:

$$T(z) \rightarrow \frac{V|b\rangle\langle b|V}{z - E_b} \quad \text{for} \quad z \rightarrow E_b. \tag{1.136}$$

The operator on the right hand side has finite rank (in this case rank 1). This is a welcome property since it maps whatever it is applied on into a linear combination of a finite number of states (here a single state). In contrast a general operator will have a continuous representation like the rest of $G(z)$, G_c, once the discrete spectrum is subtracted:

$$VG_c(z) V = \int dq \ V|q\rangle^{(+)} \frac{1}{z - E_q} {}^{(+)}\langle q|V. \tag{1.137}$$

It has become an important problem (in the context of equations describing more than 2 particles [1.9]) to find very efficient finite rank approximations of $T(z)$, since these obviously lead to an algebraic simplification.

We saw in the beginning of this section that the on-shell T-matrix obeys an unitarity relation (1.120). Now T is defined by an integral equation. As an exercise we verify the consistency of (1.127) and (1.120). Let us rewrite (1.128) as

$$T = V + TG_0 V \tag{1.138}$$

and take the adjoint

$$T^+ = V^+ + V^+ G_0^* T^+. \tag{1.139}$$

We assume V to be hermitean and subtract the two equations (1.128) and (1.139):

$$T - T^+ = VG_0 T - VG_0^* T^+ = VG_0(T - T^+) - V(G_0^* - G_0) T^+ \tag{1.140}$$

or

$$(1 - VG_0)(T - T^+) = V(G_0 - G_0^*)T^+. \tag{1.141}$$

Now according to the Lippmann-Schwinger equation (1.128) we know the inverse of $(1 - VG_0)$:

$$(1 + TG_0)(1 - VG_0) = 1. \tag{1.142}$$

Therefore we can solve (1.141) for $T - T^+$ and using (1.138) again get

$$T - T^+ = (1 + TG_0)V(G_0 - G_0^*)T^+ = T(G_0 - G_0^*)T^+. \tag{1.143}$$

Then for $z = E + i\varepsilon$ and $\varepsilon \to 0$ we end up with

$$T - T^+ = -2\pi i T\delta(E - H_0)T^+, \tag{1.144}$$

which in momentum space representation is indeed (1.120).

We shall end this section by introducing a third matrix, the K-matrix. This is advantageous both from a practical as well conceptual point of view as we shall now demonstrate. $T_{qq'}$ is complex because of the Cauchy-type singularity $(E_q + i\varepsilon - E_q)^{-1}$ occurring in the Lippmann-Schwinger equation (1.127). Let us exhibit that manifestly by using (1.61):

$$T_{qq'} = \langle q|V|q'\rangle + \int dq'' \langle q|V|q''\rangle \frac{P}{E_{q'} - E_{q''}} T_{q''q'}$$
$$- i\pi \int dq'' \langle q|V|q''\rangle \delta(E_{q'} - E_{q''}) T_{q''q'}. \tag{1.145}$$

Then one defines the K-matrix elements through the principal value kernel alone:

$$K_{qq'} = \langle q|V|q'\rangle + \int dq'' \langle q|V|q''\rangle \frac{P}{E_{q'} - E_{q''}} K_{q''q'}. \tag{1.146}$$

Obviously the reality of the V-matrix carries over to the K-matrix which is a welcome property from the computational point of view. Now we can relate the two equations (1.145) and (1.146) in the following manner. We rewrite (1.145) in the form

$$T_{qq'} - \int dq'' \langle q|V|q''\rangle \frac{P}{E_{q'} - E_{q''}} T_{q''q'}$$
$$= \langle q|V|q'\rangle - i\pi \int dq'' \langle q|V|q''\rangle \delta(E_{q'} - E_{q''}) T_{q''q'} \tag{1.147}$$

and comparing with (1.146) we recognize that the inverse of the operator acting on T on the left hand side of (1.147) when applied to V just yields K. Therefore (1.147) can be cast into the form

$$T_{qq'} = K_{qq'} - i\pi \int dq'' \, K_{qq''} \, \delta(E_{q'} - E_{q''}) \, T_{q''q'} \, . \tag{1.148}$$

It we take the external momenta q and q' to be on shell then (1.148) is a closed set connecting only on-shell T- and K-matrix elements:

$$T_{qq'} = K_{qq'} - i\pi mq \int d\hat{q}'' \, K_{qq''} \, T_{q''q'} \text{ (on shell)} \, . \tag{1.149}$$

This equation can be solved for the on-shell T-matrix once the on-shell K-matrix is given.

We shall encounter this important relation again in Chap. 2 in a partial wave representation, where it is just an algebraic equation.

Once K is given, even approximately, the resulting S-matrix is unitary. To show that let us first factor out the energy conserving δ-function in (1.105) by defining

$$S_{qq'} \equiv \frac{\delta(E_q - E_{q'})}{mq} \, \hat{S}_{qq'} \, . \tag{1.150}$$

Then (1.105) tells us the connection between the on-shell quantities:

$$\hat{S}_{qq'} = \delta(\hat{q} - \hat{q}') - 2i\pi mq \, T_{qq'} \, . \tag{1.151}$$

Using (1.149) we can express $\hat{S}_{qq'}$ in terms of $K_{qq'}$. Let us go over to a matrix notation, $\underset{\approx}{T} = \{T_{qq'}\}$ etc. Then we find easily

$$\underset{\approx}{\hat{S}} = (1 + i\pi mq \underset{\approx}{K})^{-1} (1 - i\pi mq \underset{\approx}{K}) \, , \tag{1.152}$$

which is manifestly unitary, if K is hermitean.

The hermiticity of $\underset{\approx}{K}$ follows trivially from (1.146) if $\underset{\approx}{V}$ is hermitean.

1.6 *S*-Matrix Pole Trajectories

The on-shell relation (1.151) between $\hat{S}_{qq'}$ and $T_{qq'}$ tells us that the poles of $\hat{S}$ are the same as for T. The latter ones occur, as we saw, just at the bound state energies. What will happen if we change the strength λ of the interaction, $V \equiv \lambda v$? Clearly the binding energies E_b will move and therefore the poles. Let us weaken λ such that one E_b-value goes to zero. For even smaller λ's it has to disappear from the energy plane cut along $0 \leqslant E < \infty$. The reason is, that as an eigenvalue of H it has always to be real, and for local potentials it cannot be embedded in the continuous spectrum $E \geqslant 0$ of H.

Exercise: Prove that a local potential of finite range cannot support a bound state embedded into the continuum.
Hint: Use the square integrability condition to determine the wave function outside the range of the potential.

Assume now that $T(z)$ can be continued analytically into sheets adjacent to the upper and lower rims of the cut $0 \leqslant z < \infty$. Then it makes sense to ask for the continuation of the trajectory of E_b if λ is decreased further. It is common usage to call the cut, energy plane the physical sheet since the bound state poles and the physical $T(z)$-values determining the scattering process are located there. Adjacent sheets are called nonphysical. Specifically, the one connected to the upper rim of the cut is usually called the second sheet, in contrast to the first sheet as the physical one is also called.

What physical insight can be expected if we manage to follow the path of a S-matrix pole onto the second sheet? If it stays close to the adjacent upper rim of the cut of the physical sheet, we expect it to influence the physical process. As we shall show below, this pole will produce a structure, a resonance, in the cross section.

Let us now try to find an analytic continuation of the Lippmann-Schwinger equation for $T(z)$

$$T_{qq'}(z) = V_{qq'} + \int dq'' \, V_{qq''} \frac{1}{z - E_{q''}} T_{q''q'}(z) \tag{1.153}$$

into the second sheet. We choose $z = E + i\varepsilon$, $E > 0$, $\varepsilon > 0$. Then clearly the kernel of (1.153) will have a pole at $q'' = q_0 = \sqrt{2mE} + i\varepsilon$. Let us work with spherical polar coordinates, $q \equiv q\hat{q}$. We may ask now, can one deform the path of integration in q'' in the neighbourhood of q_0 into the lower half plane such that E can be chosen on the real axis and even below in the second sheet? It is possible if $V_{qq''}$ and $T_{q''q'}$ are analytic in the required neighbourhood of the q''-axis. Let us regard $V_{qq''}$ which is given as

$$V_{q\hat{q}, q''\hat{q}''} = \frac{1}{(2\pi)^3} \int dx \, e^{-iq\hat{q}x} V(x) e^{iq''\hat{q}''x}. \tag{1.154}$$

Clearly for a finite range potential the integral continues to exist for complex q and/or q'' and it defines an analytic function in q'' and q. Moreover the right hand side of (1.153) can be considered as an integral representation of $T_{qq'}(z)$ in its dependence on q and because of the properties of V it is analytic for the required complex q-values. Therefore the deformation of the path of integration in q'' is justified and we can move E into the second sheet below the real axis. The formulation will be more transparent if we shift the path of integration back to the real axis again. Clearly in doing so, we sweep over the pole at $q'' = q_0 = \sqrt{2mE}$ and pick up a residue. The obvious result is

$$T_{qq'}^{\mathrm{II}}(E) = V_{qq'} + \int dq'' \, V_{qq''} \frac{1}{E - E_{q''}} T_{q''q'}^{\mathrm{II}}(E)$$

$$- mq_0 2\pi i \int d\hat{q}'' \, V_{qq_0\hat{q}''} T_{q_0\hat{q}''q'}^{\mathrm{II}}(E). \tag{1.155}$$

We added a superscript II to indicate that this equation is valid in the lower half plane of the second sheet. In comparison to the equation valid in the first sheet we encounter an additional term. It contains a new amplitude $T^{II}_{q_0\hat{q}''q'}(E)$, for which we require the supplementary equation

$$T^{II}_{q_0\hat{q}q'}(E) = V_{q_0\hat{q}q'} + \int dq'' V_{q_0\hat{q}q''} \frac{1}{E-E_{q''}} T^{II}_{q''q'}(E)$$
$$- m q_0 2\pi i \int d\hat{q}'' V_{q_0\hat{q}q_0\hat{q}''} T^{II}_{q_0\hat{q}''q'}(E) . \tag{1.156}$$

The two equations (1.155, 156) define the T-matrix in the second sheet. Let us now concentrate on the poles E_p of $T(E)$. In their neighbourhood $T(E)$ has the form

$$T^{II}_{qq'}(E) = \frac{R^{II}(q)}{E-E_p} , \tag{1.157}$$

which when inserted into (1.155, 156) yields right at the pole the homogeneous equations

$$R^{II}(q) = \int dq'' V_{qq''} \frac{1}{E-E_{q''}} R^{II}(q'')$$
$$- m q_0 2\pi i \int d\hat{q}'' V_{qq_0\hat{q}''} R^{II}(q_0\hat{q}'') \tag{1.158}$$

$$R^{II}(q_0\hat{q}) = \int dq'' V_{q_0\hat{q}q''} \frac{1}{E-E_{q''}} R^{II}(q'')$$
$$- m q_0 2\pi i \int d\hat{q}'' V_{q_0\hat{q}q_0\hat{q}''} R^{II}(q_0q'') . \tag{1.159}$$

Clearly the q'-dependence of the driving term in (1.155) and (1.156) is now absent.

If we go back to the first sheet for the bound state poles all the q_0-terms are absent and we simply have

$$R^{I}(q) = \int dq'' V_{qq''} \frac{1}{E-E_{q''}} R^{I}(q'') . \tag{1.160}$$

It is of great practical importance to know the types of trajectories in the second sheet, as we shall see below. We shall now assume that V is rotationally invariant. Therefore the orbital angular momentum l is conserved and we can make the ansatz

$$R^{II}(q) = Y_{lm}(q) R^{II}_l(q) . \tag{1.161}$$

Using the well known expansion of a plane wave into orbital angular momentum states

$$e^{iqx} = 4\pi \sum_{lm} Y_{lm}(x) Y_{lm}^*(\hat{q}) i^l j_l(qr) \tag{1.162}$$

and the definition (1.154) of $V_{qq'}$ it is a simple exercise to reduce (1.158, 159) to

$$R_l^{II}(q) = \int_0^\infty dq' q'^2 v_l(qq') \frac{1}{E - E_{q'}}$$
$$\times R_l^{II}(q') - mq_0 2\pi i v_l(qq_0) R_l^{II}(q_0) \quad \text{and} \tag{1.163}$$

$$R_l^{II}(q_0) = \int_0^\infty dq' q'^2 v_l(q_0q') \frac{1}{E - E_{q'}}$$
$$\times R_l^{II}(q') - mq_0 2\pi i v_l(q_0q_0) R_l^{II}(q_0), \tag{1.164}$$

where $v_l(qq')$ is given by

$$v_l(qq') = \frac{2}{\pi} \int_0^\infty dr \, r^2 j_l(qr) V(r) j_l(q'r). \tag{1.165}$$

Finally we can eliminate $R_l^{II}(q_0)$ to arrive at

$$R_l^{II}(q) = \int_0^\infty dq' q'^2 \left[v_l(qq') - v_l(qq_0) \frac{mq_0 2\pi i}{1 + mq_0 2\pi i v_l(q_0q_0)} v_l(q_0q') \right]$$
$$\times \frac{1}{E - E_{q'}} R_l^{II}(q'). \tag{1.166}$$

We see that the eigenvalue problem in the second sheet is distinguished from the one in the first sheet by an additional potential, which is complex, energy-dependent, and of finite rank. Switching to an abstract notation (1.166) reads

$$|R\rangle = v G_0 |R\rangle - |v(q_0)\rangle C(q_0) \langle v(q_0)|G_0|R\rangle. \tag{1.167}$$

Therefore the eigenvalue problem achieves the form

$$|R\rangle = -(1 - v G_0)^{-1} |v(q_0)\rangle C(q_0) \langle v(q_0)|G_0|R\rangle \quad \text{or} \tag{1.168}$$

$$\langle v(q_0)|G_0|R\rangle [1 + C(q_0)] \langle v(q_0)|G_0(1 - v G_0)^{-1}|v(q_0)\rangle = 0. \tag{1.169}$$

The nontrivial solution requires the bracket to be zero, which, again back in an explicit notation, reads

$$1 + \frac{mq_0 2\pi i}{1 + mq_0 2\pi i v_l(q_0q_0)} \int_0^\infty dq \, q^2 \int_0^\infty dq' q'^2$$

$$\times v_l(q_0q) \frac{1}{E - E_q} \langle q|(1 - v G_0)^{-1}|q'\rangle v_l(q'q_0) = 0. \tag{1.170}$$

If we multiply (1.170) by the denominator of the second term and note that

$$G_0(1 - vG_0)^{-1} = (E - H_0 - v)^{-1}$$

equation (1.170) can be written in the compact form

$$0 = 1 + mq_0 2\pi i \, t_l^I(q_0, q_0, E)$$

$$\equiv 1 + mq_0 2\pi i \left[v_l(q_0 q_0) \right.$$

$$\left. + \int_0^\infty dq \, q^2 \int_0^\infty dq' q'^2 v_l(q_0 q) \langle q | (E - H_0 - v)^{-1} | q' \rangle v_l(q' q_0) \right]. \quad (1.171)$$

Here we have introduced the partial wave t-matrix t_l [see (1.132)]. The superscript I indicates that the resolvent operator $(E - H_0 - v)^{-1}$ is evaluated in the first sheet, however the off-shell momentum q_0 belongs to the second sheet.

It is an easy numerical excercise to determine the zeros $q_0 = \sqrt{2mE}$ of (1.170). Again note that $G_0(E)$ is evaluated in the first sheet. One may start with a bound state pole as defined by (1.163) without the second term on the right hand side. Then weakening the potential strength λ the bound state pole will move towards and eventually reach $E = 0$, and then for even weaker λ-values (1.163, 164) start to hold. Typical pole trajectories for $l = 0, 1, 2$ are shown in Fig. 1.1a–c.

In each case the first and second sheet is shown. Whereas the s-wave trajectory in the second sheet moves back on the negative real axis, the trajectories for $l > 0$ leave the neighbourhood of the positive real axis with increasing real part of E. For $l > 0$, the rate at which the trajectories leave the positive real axis decreases with increasing l. As examples we have chosen cases of practical interest. The nucleon-nucleon interaction in the state $^1S_0$ (see Chap. 2) does not bind two nucleons but is strong enough to support a virtual state close to $E = 0$. A parametrisation according to *Reid* [1.10] in that state $^1S_0$ is

$$V(r) = \left(-10.463 \, \frac{e^{-\mu r}}{\mu r} - 1650.6 \, \frac{e^{-4\mu r}}{\mu r} + 6484.2 \, \frac{e^{-7\mu r}}{\mu r} \right) \text{MeV}$$

$$(\mu = 0.7 \text{ fm}^{-1}). \quad (1.172)$$

Some pole-positions for certain strength parameters λ around the physical value $\lambda = 1$ are shown in Fig. 1.1a. The nucleon-nucleus interaction can be parametrised quite successfully [1.11] by an average single-particle potential like

$$V(r) = -(V_0 + i W_0) \, \frac{1}{1 + e^{(r-R)/a}}. \quad (1.173)$$

Here V_0 and W_0 are the depth of the real and imaginary (absorptive) part and R and a are the radius and the surface width of the nucleus, respectively.

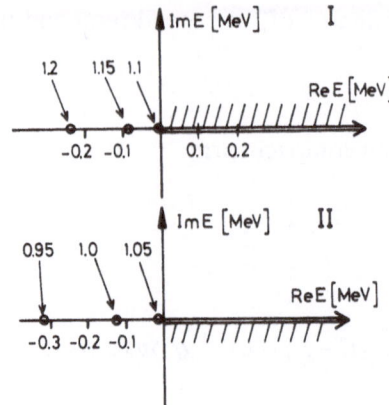

a

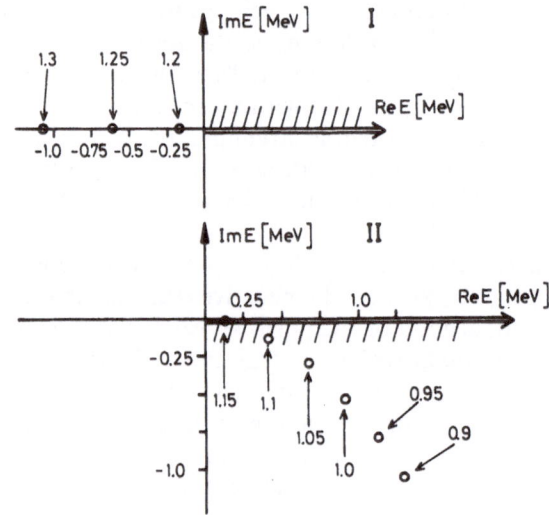

b

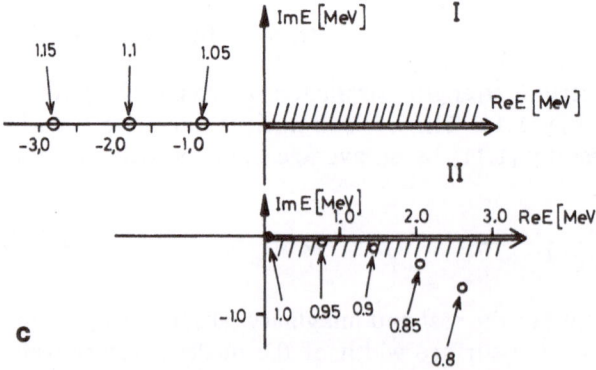

c

Fig. 1.1. (a) One S-matrix pole trajectory in the state $^1S_0$ for the two-nucleon potential (1.172) in the first and second sheet. The numbers denote the strength parameters. **(b)** One S-matrix pole trajectory in the p-state for the nucleon-nucleus interaction (1.173) in the first and second sheet. **(c)** One S-matrix pole trajectory in the d-state for the nucleon-nucleus interaction (1.173) in the first and second sheet

Figure 1.1b corresponds to a situation like $n - \alpha$ scattering ($V_0 = -40$ MeV, $R = 1.25 A^{1/3}$ fm, $A = 4$, $a = 0.65$ fm, $W_0 = 0$) and shows a low energy p-wave pole, while Fig. 1.1c describes a low-energy d-wave pole as it occurs near ^{16}O ($V_0 = -45$ MeV, $W_0 = 0$, $R = 1.25 A^{1/3}$ fm, $A = 16$, $a = 0.65$ fm). In reality of course a spin-orbit force has to be added, which leads to $P_{3/2} - P_{1/2}$ and $d_{5/2} - d_{3/2}$ splittings, respectively. In the latter case the nucleon is bound in the $d_{5/2}$ state whereas in the $d_{3/2}$ state it only interacts resonatingly with ^{16}O.

How does such a nearby pole of T situated at $E_p = E_r - i\Gamma/2$ in the second sheet influence the cross section? Taking the absolute square of T the cross section will exhibit a resonance behaviour $\propto 1/((E - E_r)^2 + \Gamma^2/4)$, where the width of the resonance is given by the imaginary part of the pole position. From the above examples we see that for a given position of the resonance, E_r, the imaginary part $-i\Gamma/2$ decreases with increasing orbital angular momentum. The resonance gets sharper. This is intuitively expected since with larger l the centrifugal barrier $l(l+1)/r^2$ increases and the particle at low energy has to tunnel through a broader and broader barrier to leave the domain D. For an infinitely high barrier we would have a bound state at a positive energy E_r and the width would be zero. For an s-state however the barrier is absent and there is no mechanism in the case of a local potential to keep the particle close to D for a positive energy, and the pole trajectory bends immediately to negative energies on the second sheet.

As a final comment we should point out that it should appear to be surprising that the pole trajectory for $l > 0$, coming from the negative real axis towards $E = 0$, bends onto the lower half plane of the second sheet, since it could equally well move upwards onto the upper half of the unphysical sheet connected to the lower rim of the cut $0 \leqslant E < \infty$. If we would have carried through the analytical continuation of the Lippmann-Schwinger equation (1.153) starting from $z = E - i\varepsilon$, $E > 0$, towards E-values with $\text{Im}\{E\} > 0$ we would have found just the imaginary conjugate of the set (1.158, 159). Therefore, there are indeed pole trajectories which are just mirror images of the trajectories shown in Fig. 1.1. A closer inspection [1.5] reveals that for $l > 0$ two poles always meet at $E = 0$, one coming from $E < 0$ in the first and one from $E < 0$ in the second sheet, respectively, and that they separate again along the resonance and "antiresonance" trajectories, respectively. Since the physical amplitude lives on the upper rim of the cut it is influenced by the nearby pole in the lower half plane of the second sheet and hardly by the "antiresonance" pole "around the corner" (the branch point $E = 0$). For $l = 0$ the trajectory does not leave the negative real axis of the first sheet or the second sheet near $E = 0$. The encounter with a partner occurs only at a certain distance away from $E = 0$ in the second sheet. The use of Jost functions in a partial-wave basis is a very powerful tool to study analytic properties of that type and is described elsewhere [1.5].

1.7 Criteria for Divergence or Convergence of the Neumann Series

Let us consider the perturbation series in V for the T-operator

$$T(z) = V + VG_0(z)V + VG_0(z)VG_0(z)V + \dots . \qquad (1.174)$$

This series, resulting from iterating the Lippmann-Schwinger equation (1.128), is called the Neumann series. In case it converges it is obviously a solution to (1.128) — but does it converge? Assume V supports a bound state. Then according to (1.136) $T(z)$ has a pole at $z = E_b < 0$. On the other hand the individual terms of the Neumann series

$$V \frac{1}{E_b - H_0} V \frac{1}{E_b - H_0} V \dots \qquad (1.175)$$

are all finite. The only way to create a pole is the divergence of the series for $z \to E_b$. How does the series behave in the neighbourhood of $z = E_b$ and at positive energies?

The key to the answer is to regard a generalisation [1.12, 1.13] of the bound state eigenvalue problem

$$|b\rangle = G_0(E_b V|b\rangle \qquad (1.176)$$

namely

$$\eta_\nu |\Gamma_\nu\rangle = G_0(E_b) V |\Gamma_\nu\rangle . \qquad (1.177)$$

Certainly (1.177) has the solution $|\Gamma\rangle = |b\rangle$ and $\eta = 1$. Furthermore every solution of (1.177) is a solution of the Schrödinger equation with the potential V/η_ν. Moreover it is a bound state, since according to (1.177) $\langle x|G_0(E_b)|x'\rangle$ is exponentially decreasing. If V is attractive, then a value $\eta_\nu < 1$ makes the potential V/η_ν stronger than V and at a certain value η_ν another bound state occurs at E_b. Obviously this can be continued and we expect a sequence of discrete eigenvalues

$$\eta_1 = 1, \eta_2 < \eta_1, \eta_3 < \eta_2, \dots . \qquad (1.178)$$

If V has attractive and repulsive parts, like the $N-N$ interaction or atom-atom interactions, there will be also negative eigenvalues, which lead to bound states from the sign reversed repulsive parts of V.

From this discussion, the occurrence of other than discrete eigenvalues would be surprising. Indeed the mathematicians tell us that the kernel of (1.177) at $E = E_b$ is of the Hilbert-Schmidt type [1.14] and therefore has only discrete eigenvalues.

An integral kernel K is of the Hilbert-Schmidt type if its Hilbert-Schmidt norm exists:

$$\|K\|_{HS}^2 \equiv \mathrm{Tr}\{KK^+\} < \infty. \tag{1.179}$$

This is easily verified for $K(z) \equiv G_0(z) V$ and $z \neq [0, \infty)$. Indeed

$$\mathrm{Tr}\{KK^+\} = \int dx \int dx' |K(x, x', z)|^2 = \int dx \int dx' |G_0(x, x', z)|^2 V^2(x')$$

$$= \int dx \int dx' \left| \frac{1}{(2\pi)^3} \int dq \, \frac{\exp[iq(x-x')]}{z-q^2/2m} \right|^2 V^2(x')$$

$$= \int dx' \frac{1}{(2\pi)^3} \frac{(2\pi m)^2}{\mathrm{Im}\{\sqrt{2mz}\}} V^2(x') \equiv \frac{c}{\mathrm{Im}\{\sqrt{2mz}\}} \|V\|^2. \tag{1.180}$$

Furthermore the eigenvalues of a Hilbert-Schmidt kernel can accumulate only at $\eta = 0$. Therefore for instance there can only be a finite number of eigenvalues located outside the unit circle. This is an important point for the control of divergence or convergence of the Neumann series, as we shall see below.

We now study the property of $\eta_v(z)$ as a function of z. Let us start the discussion by considering negative energies $z < 0$. Then the eigenvalues $\eta_v(z)$ are real.

Exercise: Prove the reality of $\eta_v(z)$ for $z < 0$
Hint: see [1.13].

Furthermore they increase in magnitude if z increases. This property is obvious since the potential strength $1/\eta_v$ has to decrease if the bound state energy $z < 0$ increases. Clearly for $z \to -\infty$ all the η's tend towards zero.

Now let us consider positive energies at which scattering takes place. The eigenvalue problem defined in (1.177) for $z < 0$ can obviously be generalized immediately for the upper and lower half z-plane without any problem, since $\sqrt{2mz}$, which controls the behaviour of $G_0(x, x', z)$, as given in (1.77), will always have a positive imaginary part and thus will lead to a square integrable eigenstate $|\Gamma\rangle$. Therefore $\eta_v(z)$ is an analytic function in the z-plane, cut along $0 \leqslant z < \infty$. The limits on the upper and lower rim of the cut are defined as

$$\lim_{\varepsilon \to 0} G_0(E \pm i\varepsilon) V |\Gamma_v^{(\pm)}(E)\rangle = \eta_v^{(\pm)}(E) |\Gamma_v^{(\pm)}(E)\rangle. \tag{1.181}$$

According to (1.77) the eigenstates $\langle x | \Gamma_v^{(\pm)} \rangle$ will no longer be square integrable, but purely outgoing $(+)$ or purely incoming $(-)$ for $|x| \to \infty$. Moreover, they will be regular at the origin, and of course they will be solutions to the Schrödinger equation with the potential $V/\eta_v^{(\pm)}(E)$. For a real (hermitean) potential, however, the flux violating asymptotic behaviour $\exp(\pm i\sqrt{2mE}\,x)/x$ of $\langle x\,|\,\Gamma_\nu^{(\pm)}\rangle$ is not possible. Thus the eigenvalues $\eta_\nu^{(\pm)}(E)$, $E>0$, have to be complex.

Let us summarize. Following z along the real axis from negative to positive values, the $\eta_\nu(z)$ are real for $z<0$ and increase in magnitude up to the point $z=0$, where we encounter a branch point. As z increases further, $\eta_\nu(z)$ will acquire an imaginary part. It is $\eta_\nu^{(+)}(E)$ at the upper rim and $\eta_\nu^{(-)}(E) = (\eta_\nu^{(+)}(E))^*$ at the lower rim. For $E\to\infty$ the trajectories traced out by $\eta_\nu^{(\pm)}(E)$ have to go back into the unit circle (if they are outside at all), since in that limit the kernel $K(E)$ obviously decreases in magnitude. Again it is an easy numerical exercise to calculate the trajectories $\eta_\nu(z)$. We show in Fig. 1.2a, b a few of them belonging to the potentials of Fig. 1.1a, c, respectively. The nucleon-nucleon potential (1.172) (valid in the state $^1S_0$, see Sect. 2.6) does not support a bound state, but it supports a virtual state near $E=0$. This

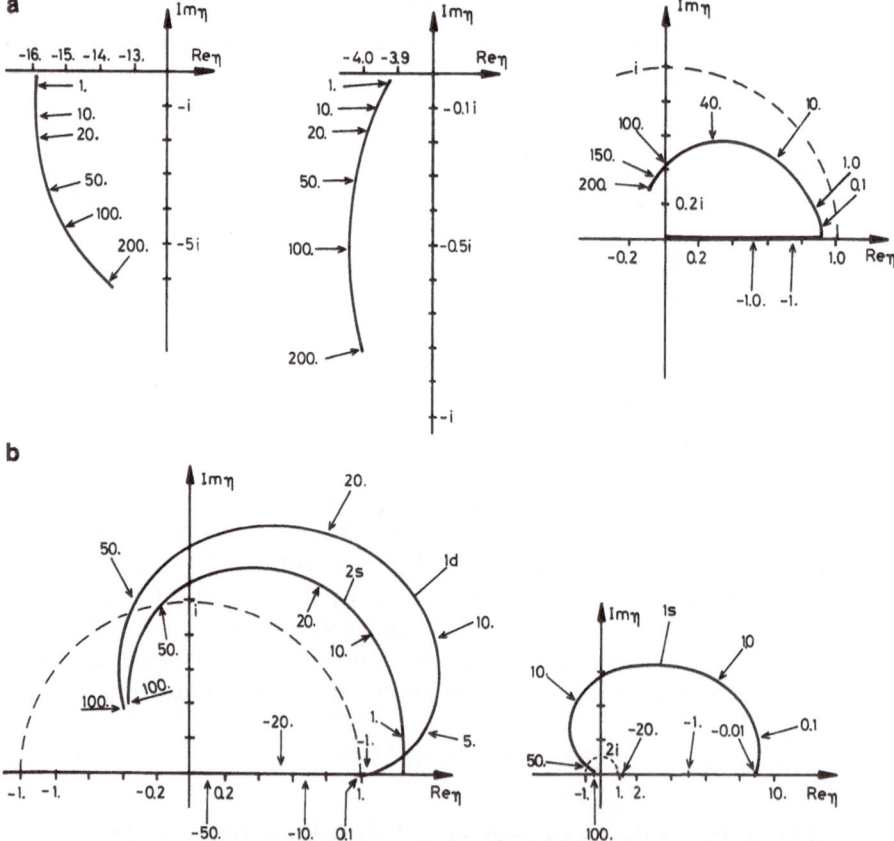

Fig. 1.2. (a) The first three eigenvalues η_ν largest in magnitude for the two-nucleon interaction (1.172) in the state $^1S_0$. The numbers indicate the energies in MeV. **(b)** The largest eigenvalues in the d- and s-state for the nucleon-nucleus interaction (1.173)

is shown in Fig. 1.2a, as reflected by the fact that the largest positive eigenvalue does not reach $\eta = 1$ before leaving the real axis at $E = 0$. For this potential there are also strong negative eigenvalues. For negative energies (not
shown in Fig. 1.2a) they are linked to bound states in the sign reversed poten-
tial (1.172), whose magnitude is modified by $|1/\eta|$. As can be seen in Fig.
1.2a, there are some eigenvalues, which stay outside the unit circle up to very
high energies, and the Neumann series would converge only at energies for
which that potential picture is already meaningless. In Fig. 1.2b we see that
the underlying potential supports two s-wave bound states. For energies large
with respect to the potential depth the two trajectories return into the unit
circle. The d-state is not bound since the trajectory leaves the real axis just before reaching 1. This is not visible on the scale of the Fig. 1.2b. The remaining
eigenvalues all stay within the unit circle and are not shown.

What can we conclude from this insight? Let us apply the Neumann series
(1.174) onto an eigenstate $|\Gamma_\nu\rangle$. Then due to (1.177), which we assume to be
written for an arbitrary z, we get

$$T(z)|\Gamma_\nu(z)\rangle = V|\Gamma_\nu(z)\rangle + V\eta_\nu(z)|\Gamma_\nu(z)\rangle + \cdots$$
$$= V|\Gamma_\nu(z)\rangle(1 + \eta_\nu(z) + \eta_\nu^2(z) + \cdots). \tag{1.182}$$

This sum diverges if $|\eta_\nu(z)| > 1$. In general the Neumann series for $T(z)$ is
applied to a state which has components with respect to all $|\Gamma_\nu(z)\rangle$. In that
case the Neumann series will diverge whenever there is at least one eigenvalue
larger in magnitude than 1. It will converge only if all eigenvalues are inside
the unit circle. This is a necessary condition for convergence and it is sufficient
as will be shown in Sect. 2.7c. Now we know that the mere existence of a
bound state has the consequence that at least one eigenvalue will have a magnitude larger than 1 for $z > E_b$. Therefore the Neumann series applied on a
general state will diverge. Only if the energy is large enough, will that eigenvalue return into the unit circle and the kernel be weak enough to be iterated in
a convergent manner.

If we regard Fig. 1.2b for the $1d$-state, we see that the eigenvalue η_d is very
close to 1 near $E \approx 0$ MeV. This is the energy at which we encountered the resonance in the d-state according to Fig. 1.1c. The reason is, as we shall show
below, that $\eta_\nu(z)$ will become exactly 1 as it is analytically continued towards
$z = E_p$ in the second sheet. If the resonance position E_p is close to the real axis
we can expect that η will stay near the value 1 at neighbouring real energies E.
This is indeed the case in the above example. We conclude that in the case of a
resonance, we encounter on the real energy axis a complex eigenvalue which is
in the neighbourhood of 1 and the Neumann series will diverge or converge according to whether η is just inside or still outside the unit circle.

It remains to justify the analytic continuation of $\eta(z)$ onto the second
sheet, especially towards the pole position of $T(z)$. Comparing (1.160, 176)
and putting

$$|R\rangle = V|b\rangle \tag{1.183}$$

we see that they are of course equal. Therefore due to (1.177), the position of the bound state pole of $T(z)$ can be defined implicitly as

$$\eta(\lambda, E_b) = 1, \tag{1.184}$$

where λ plays again the role of the varying strength parameter of V, which we assume to be real. Let us now continue analytically (1.177), again written for a complex z, onto the second sheet. If we switch to the new state $|R_v\rangle \equiv V|\Gamma_v\rangle$ the eigenvalue problem to be continued,

$$\eta_v(z)|R_v(z)\rangle = VG_0(z)|R_v(z)\rangle \tag{1.185}$$

is identical to the one regarded in the previous section, where however η was kept equal to 1. Therefore $\eta_v(\lambda, z)$ is defined on the second sheet for a finite range potential, and $\eta_v(\lambda, z) = 1$ defines the position of a pole of $T(z)$ for a certain strength parameter λ.

Summarizing we find that in interesting cases where V supports bound states or resonances, the Neumann series will diverge, at least for low energies. We shall see in Sect. 2.7 how that divergent series can nevertheless be summed up to the correct solution using the Padé technique. Of course one can always solve the integral equation (1.128) directly, as we shall discuss in some details in Sect. 2.7.

The loss of the Hilbert-Schmidt property of the kernel for $E > 0$ as expressed in (1.180) can be avoided by the trick of using the kernel $V^{1/2}G_0V^{1/2}$. Then for scattering energies $E > 0$ a rigorous justification for the convergence or divergence property of the Neumann series can be formulated [1.15] within the context of standard functional analysis.

Let us end this section with a remark on a finite rank approximation of the kernel $K = VG_0$ of the Neumann series. This is of interest for formal developments in reaction theory and also to some extent in practice. We recognised that the discrete structures created by the potential, bound states or resonances, cause the divergence of the Neumann series in K, or in other words make the kernel large. The parts of K linked to these structures can be extracted in the following manner [1.13]. Consider the separable kernel

$$K_s(z) \equiv \eta_v(z)|\Gamma_v(z)\rangle \frac{1}{\langle\Gamma_v(z^*)|V|\Gamma_v(z)\rangle} \langle\Gamma_v(z^*)|V \tag{1.186}$$

and look for the eigenvalues of $K' \equiv K - K_s$:

$$K'|\Theta'_\mu\rangle = \eta'_\mu|\Theta'_\mu\rangle. \tag{1.187}$$

The claim is that the eigenvalues and eigenvectors η'_μ and $|\Theta'_\mu\rangle$ are identical to η_μ and $|\Gamma_\mu\rangle$ with the exception of the eigenvalue to $|\Theta_v\rangle = |\Gamma_v\rangle$, which is zero.

Therefore if η_ν was a large eigenvalue for K it will be a "small" one for K'. This can be generalised by replacing K_s by a finite sum of terms of this type. Then in K' the corresponding eigenvalues will be reduced to zero. Clearly this can be used to collect the finite number of eigenstates of K with eigenvalues outside the unit circle into a finite rank approximation K_s of K, and to treat the remaining kernel K' as a perturbation. The perturbation treatment is justified since K' is small enough to lead to a convergent Neumann series by its very construction. One proceeds according to the following often encountered pattern.

$$T = V + KT = V + K'T + K_s T. \qquad (1.188)$$

The small part is explicitly inverted:

$$T = (1 - K')^{-1}(V + K_s T). \qquad (1.189)$$

In our case K_s has the driving term V on the left, $K_s \equiv VM_s$. Therefore (1.189) can be written as

$$T = T' + T'M_s T \qquad (1.190)$$

with T' given through

$$T' = V + K'T'. \qquad (1.191)$$

The driving term T' in (1.190) is assumed to be calculable in a low order iteration. Due to the finite rank nature of M_s (1.190) can be solved purely algebraically.

The claims following (1.187) are based on the orthogonality and completeness properties of the Weinberg eigenstates. The interested reader is refered to the very clear presentation in the original work [1.13].

2. Scattering Theory for the Two-Nucleon System

The scattering of two particles upon each other serves as the simplest example for the methods of the previous section. Applying them to two nucleons adds the discrete degrees of freedom of spin 1/2-particles. The inclusion of spin enriches the system considerably. The many possibilities of flipping the spins of the two colliding nucleons yield sensitive observables for studying the nucleon-nucleon force [2.1]. We shall present a self-contained description including very concrete methods of calculating the various spin observables.

2.1 Density Matrices for the Initial and Final State

The nucleon in the beam or target can live in two spin states $|m\rangle$, $m = \pm 1/2$, which characterise the two values of the spin-component with respect to a certain direction. Thus for the two nucleons, a complete orthonormal basis in spin space is

$$|\lambda_i\rangle \equiv \{|m_1\rangle|m_2\rangle\}, \tag{2.1}$$

where $i = 1, 2, 3, 4$ numerates the 4 possible states. A general pure spin state is

$$|n\rangle = \sum_i a_i^{(n)}|\lambda_i\rangle. \tag{2.2}$$

Whereas we have assumed up to now (and shall continue to do so) that the momentum distribution for the particles is described by a pure state, in other words there is only one type of wave packet, the spin states of the nucleons in beam and target are in general in a mixed state, where the states $|n\rangle$ occur with the probabilities p_n. Then according to standard rules [2.2], the expectation value of a spin observable $\hat{O}$ is

$$\langle \hat{O} \rangle = \frac{\sum\limits_n p_n \langle n|\hat{O}|n\rangle}{\sum\limits_n p_n \langle n|n\rangle} = \frac{\mathrm{Tr}\{\hat{\varrho}\hat{O}\}}{\mathrm{Tr}\{\hat{\varrho}\}}. \tag{2.3}$$

In the last equality we introduced the density operator

$$\hat{\varrho} \equiv \sum_n |n\rangle p_n \langle n|. \tag{2.4}$$

In the basis of (2.1) Eqs. (2.4, 3) read

$$\hat{\varrho} = \sum |\lambda_j\rangle \varrho_{ji} \langle \lambda_i| \quad \text{with} \tag{2.5}$$

$$\varrho_{ji} = \sum_n a_j^{(n)} p_n a_i^{(n)*} \quad \text{and} \tag{2.6}$$

$$\langle \hat{O} \rangle = \frac{\sum \varrho_{ji} O_{ij}}{\sum \varrho_{ii}} \equiv \frac{\mathrm{Tr}\{\varrho O\}}{\mathrm{Tr}\{\varrho\}} \quad \text{with} \tag{2.7}$$

$$O_{ji} = \langle \lambda_j | \hat{O} | \lambda_i \rangle. \tag{2.8}$$

The unsubscripted quantities ϱ and O denote the matrices (2.6, 8).

How do we know which density operator corresponds to a certain experimental situation of beam and target? According to (2.5) $\hat{\varrho}$ depends on 16 real numbers, since ϱ is a 4 dimensional hermitean matrix. They can be related to the expectation values of 16 linearily independent hermitean matrices S^μ, $\mu = 1, \ldots, 16$. A convenient set is built up by the complete basis of 2×2 matrices $\{1, \sigma_x, \sigma_y, \sigma_z\}$ chosen for both particles, where the σ's are the standard Pauli matrices. Let us introduce the notation $\sigma_0 \equiv 1$ in addition to $\sigma_1 \equiv \sigma_x$, $\sigma_2 \equiv \sigma_y$, $\sigma_3 \equiv \sigma_z$, then

$$\{S^\mu\} = \{\sigma_\alpha^{(1)} \otimes \sigma_\beta^{(2)}\}, \quad \alpha, \beta = 0, 1, 2, 3. \tag{2.9}$$

The set (2.9) fulfills the orthogonality condition

$$\mathrm{Tr}\{S^\mu S^\nu\} = 4\delta_{\mu\nu}. \tag{2.10}$$

Exercise: Verify (2.10).

We decompose the matrix ϱ into the set (2.9) and use (2.10) to get

$$\varrho = \sum_{\mu=1}^{16} S^\mu A_\mu = \frac{1}{4} \sum_{\mu=1}^{16} S^\mu \mathrm{Tr}\{\varrho S^\mu\}. \tag{2.11}$$

In terms of the expectation values $\langle S^\mu \rangle$ defined by (2.7) this reads

$$\varrho = \tfrac{1}{4} \mathrm{Tr}\{\varrho\} \sum_\mu S^\mu \langle S^\mu \rangle. \tag{2.12}$$

Thus the 16 expectation values in the initial state determine the initial density matrix up to the normalisation. We shall denote it by ϱ_i. Various examples will be given later.

In the two-nucleon collision, the interaction will in general change the spin states and therefore the density matrix. Obviously a two-body problem, after eliminating the force free center of mass motion, reduces to the motion of a ficticious particle with reduced mass μ in a potential. For equal mass particles $\mu = \frac{1}{2}m$. If k_1 and k_2 are the individual particle momenta, the momentum governing the relative motion is $q = (k_1 - k_2)/2$ and will also undergo in general a change during the collision, whereas the total momentum $K = k_1 + k_2$ will be conserved. The latter one will be dropped in all that follows. The stationary scattering state of relative motion obeys the Lippmann-Schwinger equation (1.34). Now, however, we have to specify also the initial spin state. Since a mixed state is an incoherent superposition of pure states which undergo separately the scattering process, one has first to determine the scattering state $|qn\rangle^{(+)}$ developing out of a pure initial spin state $|n\rangle$ and initial relative momentum q. Thus the Lippmann-Schwinger equation is generalised to

$$|qn\rangle^{(+)} = |q\rangle|n\rangle + \sum_i |\lambda_i\rangle\langle\lambda_i|G_0^{(+)}V|qn\rangle^{(+)}. \tag{2.13}$$

Note that the particle in the state $|qn\rangle^{(+)}$ now lives in two spaces, the ordinary one and the spin space, which we made explicit by inserting the unit operator in spin space in front of $G_0 V$. The situation after scattering is determined by the asymptotic form of the wave function. Let x be the conjugate coordinate to q, $x = x_1 - x_2$. Then according to (1.90) we get for $|x| \to \infty$

$$\langle x|qn\rangle^{(+)} \to \frac{1}{(2\pi)^{3/2}}\left(e^{iqx}|n\rangle + \sum_i |\lambda_i\rangle \frac{e^{iqx}}{x}f_i(q',q)\right), \tag{2.14}$$

where

$$f_i(q',q) = -\mu(2\pi)^2\langle\lambda_i|\langle q'|V|qn\rangle^{(+)} \tag{2.15}$$

and $q' \equiv q\hat{x}$ points into the direction of observation. Since V is assumed to be spin dependent, f_i will depend in a nontrivial dynamical manner on i. The dependence of the scattering amplitude on the intial spin state can be exhibited explicitly by introducing the T-operator. According to (1.34) and (1.126) this amounts to

$$V|qn\rangle^{(+)} \equiv T(E_q)|q\rangle|n\rangle \tag{2.16}$$

and we get

$$f_i(q',q) = -\mu(2\pi)^2 \sum_j \langle\lambda_i|\langle q'|T|q\rangle|\lambda_j\rangle a_j^{(n)} \equiv \sum_j M_{ji}(q',q)a_j^{(n)}, \tag{2.17}$$

which shows explicitly the linear dependence on the initial spin coefficients $a_j^{(n)}$. The 4-dimensional M-matrix in spin space contains all the dynamical information of the scattering process.

Later we shall modify the definition of M to account for identical particles using the isospin concept.

Due to (2.14, 17), the spin state after the scattering process is given as

$$|f^{(n)}\rangle = \sum_i |\lambda_i\rangle \sum_j M_{ij}(q'q)a_j^{(n)} \equiv \sum_i |\lambda_i\rangle (a_f)_i^{(n)} \,. \tag{2.18}$$

Consequently the density matrix for the final state is

$$(\varrho_f)_{ji} = \sum_n (a_f)_j^{(n)} p_n (a_f)_i^{(n)*} = \sum_n \sum_{kl} M_{jk}(q'q)a_k^{(n)} p_n M_{il}^*(q'q)a_l^{(n)*}$$

$$= \sum_{kl} M_{jk}(q'q)(\varrho_i)_{kl} M_{li}^+(q'q) \tag{2.19}$$

or in matrix notation

$$\varrho_f = M(q'q)\varrho_i M^+(q'q) \,. \tag{2.20}$$

Note that in the first line of (2.19) we summed over the pure initial states $|n\rangle$, weighted according to the probability p_n of finding them in the beam. Once the density matrix ϱ_f is known, the expectation value of an arbitrary spin observable in the final state is

$$\langle \hat{O} \rangle_f = \frac{\text{Tr}\{\varrho_f O\}}{\text{Tr}\{\varrho_f\}} = \frac{\text{Tr}\{M\varrho_i M^+ O\}}{\text{Tr}\{M\varrho_i M^+\}} \,. \tag{2.21}$$

2.2 The General Spin Observable

The simplest observation in the final state avoids the measurement of the spin orientations. This is just the differential cross section summed over the spin orientations in the final state:

$$I \equiv \frac{\overline{d\sigma}}{d\Omega} = \frac{\sum_n p_n \sum_i \left| \sum_j M_{ij} a_j^{(n)} \right|^2}{\sum_n p_n \sum_j |a_j^{(n)}|^2}$$

$$= \frac{\sum_{ijl} M_{ij} \sum_n a_j^{(n)} p_n a_l^{(n)*} M_{li}^+}{\sum_j \sum_n a_j^{(n)} p_n a_j^{(n)*}}$$

$$= \frac{\text{Tr}\{M\varrho_i M^+\}}{\text{Tr}\{\varrho_i\}} = \frac{\text{Tr}\{\varrho_f\}}{\text{Tr}\{\varrho_i\}} \,. \tag{2.22}$$

In the first line we averaged over the distribution of the pure initial spin-states $|n\rangle$.

The most general observable links arbitrary spin orientations for both nucleons in the initial to arbitrary ones in the final state. We have the total information available, once we know the 16 expectation values $\langle S^\mu \rangle_f$ with respect to the final density matrix ϱ_f:

$$\langle S^\mu \rangle_f = \frac{\text{Tr}\{\varrho_f S^\mu\}}{\text{Tr}\{\varrho_f\}} \ . \tag{2.23}$$

We insert (2.20) and the explicit form (2.12) for ϱ_i and get

$$\langle S^\mu \rangle_f = \frac{\text{Tr}\{M\varrho_i M^+ S^\mu\}}{\text{Tr}\{M\varrho_i M^+\}} = \frac{1}{4}\text{Tr}\{\varrho_i\}\frac{\sum_\nu \langle S^\nu \rangle_i \text{Tr}\{MS^\nu M^+ S^\mu\}}{\text{Tr}\{M\varrho_i M^+\}} \tag{2.24}$$

or using (2.22)

$$\langle S^\mu \rangle_f I = \tfrac{1}{4} \sum_\nu \langle S^\nu \rangle_i \text{Tr}\{MS^\nu M^+ S^\mu\} \ . \tag{2.25}$$

Note this includes the case that no spin is measured in the final state. By choosing $S^\mu = \sigma_0^{(1)} \otimes \sigma_0^{(2)}$ (2.25) obviously reduces to (2.22). Now (2.25) reads explicitly

$$I\langle \sigma_\mu^{(1)} \sigma_\nu^{(2)} \rangle_f = \tfrac{1}{4} \sum_{\lambda\varrho} \langle \sigma_\lambda^{(1)} \sigma_\varrho^{(2)} \rangle_i \text{Tr}\{M\sigma_\lambda^{(1)}\sigma_\varrho^{(2)}M^+ \sigma_\mu^{(1)}\sigma_\nu^{(2)}\} \ . \tag{2.26}$$

This expression links the various spin expectation values in the initial state to all possible spin expectation values in the final state. Clearly the total number of possibilities including no spin measurement (choose σ_0) is $16 \times 16 = 256$. This requires at first sight, a very large number of experiments for each scattering angle and center-of-mass-energy. There are however invariance requirements, as we shall see, which forbids many transitions. This is reflected in the fact that M can be characterized by a small number of parameters, which moreover induces strong correlations between the surviving nonzero observables.

2.3 The Wolfenstein Parametrisation of the Scattering Amplitude

In nuclear physics the isospin concept is very useful. Though being only an approximate symmetry [2.3], broken on the level of electromagnetic interactions, it adds a further classification of states and correlates otherwise unlinked processes. Neuron and proton are considered to be the two magnetic states $|m_t\rangle$, $m_t = 1/2$ and $-1/2$, respectively, of an isospin $t = 1/2$ particle, the nucleon. Therefore for two nucleons the physical state $|m_{t_1} m_{t_2}\rangle$ is in general a superposition of total isospin $t = 0$ and $t = 1$:

$$|m_{t_1} m_{t_2}\rangle = \sum_t C(\tfrac{1}{2}\tfrac{1}{2}t, m_{t_1} m_{t_2} m_t)|t m_t\rangle. \tag{2.27}$$

If the two-nucleon interaction V conserves isospin, the scattering states for $t = 0$ and $t = 1$ can be treated separately. In other words the M-matrix elements can be classified according to t, $M^{t m_t}$, and the physical matrix elements will be built up according to (2.27) by linear combinations of $M^{t=0}$ and $M^{t=1}$. For the (nn) and (pp) system, the isospin is $t = 1$ and $M^{t=1}$ is already the physical amplitude. Only for the (pn) system is relation (2.27) nontrivial:

$$|pn\rangle \equiv |-\tfrac{1}{2}\tfrac{1}{2}\rangle = \frac{1}{\sqrt{2}}(|t = 0, m_t = 0\rangle + |t = 1, m_t = 0\rangle). \tag{2.28}$$

The physical amplitude is

$$M_{pn \to pn} = \tfrac{1}{2}(M^{t=1} + M^{t=0}). \tag{2.29}$$

Adding the isospin label to the spin and position (or momentum) labels of a nucleon, we can treat neutrons and protons as identical particles. Therefore the two-nucleon states have to be antisymmetric under exchange of the two particles. Clearly the exchange has to take place in all three spaces in which the state is defined: in normal space, in spin space and isospin space. As an exercise we pose the question whether isospin conservation in the two-nucleon system leads to spin conservation assuming parity invariance. The answer is easily found noting that the parity operation changes the relative position $x \equiv x_1 - x_2$ into $-x$ which is the same result achieved by permuting the particles. Now in the case of parity invariance the two nucleon states can be classified into states of good parity. Further, the isospin states for two nucleons of $t = 0$ and $t = 1$ are clearly antisymmetric and symmetric respectively, under exchange of the two particles. The same is true for the spin states of $s = 0$ and $s = 1$. Therefore an antisymmetric two-nucleon state could be, for instance, a state of positive parity (symmetric under exchange) and a spin-singlet combined with an isospin-triplet state. Since the symmetry of the space part cannot change assuming parity invariance, isospin invariance clearly entails spin conservation.

Now we can introduce the antisymmetric free state

$$\{|q\rangle|n\rangle|t m_t\rangle\}_a \equiv (1 - P_{12})|q\rangle|n\rangle|t m_t\rangle, \tag{2.30}$$

where P_{12} permutes the particles in all three spaces. Then the antisymmetric scattering state defined by

$$|q n t m_t\rangle_a^{(+)} \equiv (1 - P_{12})|q n t m_t\rangle^{(+)} \tag{2.31}$$

will obviously obey the Lippmann-Schwinger equation

$$|q n t m_t\rangle_a^{(+)} = |q n t m_t\rangle_a + \lim_{\varepsilon \to 0} \sum_{t'm_{t'}} |t'm_{t'}\rangle\langle t'm_{t'}| \sum_i |\lambda_i\rangle\langle\lambda_i|G_0 V|q n t m_t\rangle_a^{(+)}.$$

$$(2.32)$$

Consequently the scattering amplitude will be modified to

$$f_i(q'q) = -\mu(2\pi)^2 \sum_j \langle\lambda_i|\langle t m_t|\langle q'|T\{|q\rangle|\lambda_j\rangle|t m_t\rangle\}_a a_j^{(n)}.$$

$$(2.33)$$

Note that we explicitly assumed t-conservation. Otherwise the sum over t' in (2.32) could not be reduced just to the one term $|tm_t\rangle$ and in general scattering amplitudes for both t' values would appear. The m_t-value of course cannot change since it measures the charge of the two nucleons. In this approximation of isospin invariance, the M-matrix for identical nucleons is therefore given by

$$M_{m_1'm_2'm_1 m_2}^t(q'q) = -\mu(2\pi)^2\langle q'|\langle m_1' m_2'|\langle t m_t|T\{|t m_t\rangle|m_1 m_2\rangle|q\rangle\}_a.$$

$$(2.34)$$

Now we can pose the main question of this section: how can the dependence on q and q', m_1, m_2, m_1', m_2' be parametrised? For the orientation in space, one introduces three unit vectors constructed out of q and q':

$$\hat{K} \equiv (q'-q)/|q'-q|$$
$$\hat{P} \equiv (q+q')/|q+q'|$$
$$\hat{N} \equiv (q \times q')/|q \times q'|.$$

$$(2.35)$$

$\hat{K}$ and $\hat{P}$ lie in the scattering plane to which $\hat{N}$ is the normal. They can be considered as the unit vectors for a right-handed, orthogonal coordinate system. The dependence of M on the spin components $m_1 m_2 m_1' m_2'$ can be exhausted by a linear combination of the 4-dimensional matrices S^μ defined in (2.9). What do we know about that linear combination? The first requirement we shall impose is rotational invariance of V. If V is just a local spin independent interaction, as used in Chap. 1, $\langle q'|V|q\rangle$ can depend only on the scalars qq, $q'q'$, and qq'. This is obvious from (1.154). If V is spin dependent, additional scalars built with the aid of $\sigma^{(1)}$ and $\sigma^{(2)}$ will show up. The rotational invariance of V carries over immediately to the T-operator as is explicit in (1.132). Therefore the M-matrix can depend only on scalars built out of $\sigma^{(1)}$ and $\sigma^{(2)}$ contained in the S^μ's and q and q'. Let us make a table for all of them (besides the trivial constant)

$\sigma^{(i)}\hat{K}$	$\sigma^{(i)}\hat{N}$	$\sigma^{(i)}\hat{P}$	$i = 1, 2$
$\sigma^{(1)}\times\sigma^{(2)}\hat{K}$	$\sigma^{(1)}\times\sigma^{(2)}\hat{N}$	$\sigma^{(1)}\times\sigma^{(2)}\hat{P}$	
$(\sigma^{(1)}\hat{K})(\sigma^{(2)}\hat{K})$	$(\sigma^{(1)}\hat{K})(\sigma^{(2)}\hat{N})$	$(\sigma^{(1)}\hat{K})(\sigma^{(2)}\hat{P})$	(2.36)

$$(\sigma^{(1)}\hat{N})(\sigma^{(2)}\hat{K}) \qquad (\sigma^{(1)}\hat{N})(\sigma^{(2)}\hat{N}) \qquad (\sigma^{(1)}\hat{N})(\sigma^{(2)}\hat{P})$$

$$(\sigma^{(1)}\hat{P})(\sigma^{(2)}\hat{K}) \qquad (\sigma^{(1)}\hat{P})(\sigma^{(2)}\hat{N}) \qquad (\sigma^{(1)}\hat{P})(\sigma^{(2)}\hat{P}).$$

Another possible scalar $\sigma^{(1)}\sigma^{(2)}$ is implicitly included in the above group, since

$$\sigma^{(1)}\sigma^{(2)} = (\sigma^{(1)}\hat{K})(\sigma^{(2)}\hat{K}) + (\sigma^{(1)}\hat{N})(\sigma^{(2)}\hat{N}) + (\sigma^{(1)}\hat{P})(\sigma^{(2)}\hat{P}). \tag{2.37}$$

Now the matrix M has to be a linear combination of all terms in (2.36) with coefficients which depend on the scalars qq', qq and $q'q'$. The last two, however, are related to the center-of-mass energy $E = q^2/2\mu = q'^2/2\mu$ for the on-shell amplitude under discussion. Therefore the coefficients are functions of E and the center-of-mass scattering angle ϑ given by

$$\cos\vartheta \equiv \hat{q}\hat{q}'. \tag{2.38}$$

For most purposes this is still too general since the nuclear interaction is invariant under parity and time reversal operations, at least to a very high accuracy [2.4]. Under a parity operation

$$\hat{K} \rightarrow -\hat{K}$$
$$\hat{P} \rightarrow -\hat{P}$$
$$\hat{N} \rightarrow \hat{N}$$
$$\sigma^{(i)} \rightarrow \sigma^{(i)}. \tag{2.39}$$

The underlined terms in (2.36) do not remain invariant under this replacement and are therefore forbidden.

The exploitation of the time reversal invariance of V needs some more consideration. As in Sect. 1.5 we write

$$\langle q'm_1'm_2'|T|qm_1m_2 \rangle = \langle \mathscr{T}q'm_1'm_2'|\mathscr{T}T|qm_1m_2 \rangle^*$$
$$= \langle \mathscr{T}qm_1m_2|T|\mathscr{T}q'm_1'm_2' \rangle. \tag{2.40}$$

In the second equality we used the fact that $\mathscr{T}$ acting on T changes $+i\varepsilon$ into $-i\varepsilon$ in the resolvent operator G which is converted back to $+i\varepsilon$ by applying T to the left. Now in the usual phase convention [2.5], the $\mathscr{T}$-operation on an angular momentum eigenstate is

$$\mathscr{T}|jm \rangle = (-)^{j-m}|j-m \rangle \tag{2.41}$$

and (2.40) yields

$$\langle q'm_1'm_2'|T|q\,m_1m_2\rangle = (-)^{1/2-m_1+1/2-m_2}\langle -q-m_1-m_2|T|-q'-m_1'-m_2'\rangle$$
$$(-)^{1/2-m_1'+1/2-m_2'} \tag{2.42}$$

or

$$(-)^{1/2-m_1+1/2-m_2+1/2-m_1'+1/2-m_2'}M_{-m_1-m_2-m_1'-m_2'}(-q,-q')$$
$$= M_{m_1'm_2'm_1m_2}(q'q) . \tag{2.43}$$

This is the basic relation between time reversed processes from which the familiar detailed balance property [2.3] of cross sections follows.

How is the left hand side of (2.43) to be interpreted in terms of the σ-matrices which we use to expand the M-matrix? As an example, instead of

$$\sigma^{(1)}_{m_1'm_1} \equiv \langle m_1'|\sigma^{(1)}|m_1\rangle \tag{2.44}$$

we encounter on the left hand side

$$(-)^{1/2-m_1'+1/2-m_1}\langle -m_1|\sigma^{(1)}|-m_1'\rangle = \langle \mathcal{T}\tfrac{1}{2}m_1|\sigma^{(1)}|\mathcal{T}\tfrac{1}{2}m_1'\rangle. \tag{2.45}$$

Now σ as a spin operator will change sign under time reversal, therefore

$$\langle \mathcal{T}\tfrac{1}{2}m_1|\sigma^{(1)}|\mathcal{T}\tfrac{1}{2}m_1'\rangle$$
$$= -\langle \mathcal{T}\tfrac{1}{2}m_1|\mathcal{T}\sigma^{(1)}|\tfrac{1}{2}m_1'\rangle = -\langle \tfrac{1}{2}m_1|\sigma^{(1)}|\tfrac{1}{2}m_1'\rangle^*$$
$$= -\langle \tfrac{1}{2}m_1'|\sigma^{(1)}|\tfrac{1}{2}m_1\rangle = -\sigma^{(1)}_{m_1'm_1} . \tag{2.46}$$

We end up with the recipe that on the left hand side of (2.43) the σ-matrices have to be replaced by their negatives:

$$\sigma^{(i)} \to -\sigma^{(i)} . \tag{2.47}$$

Finally, since $q \to -q'$ and $q' \to -q$, it follows that

$$\hat{K} \to \hat{K}$$
$$\hat{N} \to -\hat{N} \tag{2.48}$$
$$\hat{P} \to -\hat{P} .$$

Requiring time reversal invariance therefore eliminates the further terms in (2.36) which are underlined by a dotted line.

Having built in invariance under rotation, parity- and time reversal transformations, we find the following representation of the M-matrix [2.6]:

$$M = a + b(\sigma^{(1)} - \sigma^{(2)})\hat{N} + c(\sigma^{(1)} + \sigma^{(2)})\hat{N} + m(\sigma^{(1)}\hat{N})(\sigma^{(2)}\hat{N})$$
$$+ (g+h)(\sigma^{(1)}\hat{P})(\sigma^{(2)}\hat{P}) + (g-h)(\sigma^{(1)}\hat{K})(\sigma^{(2)}\hat{K}) . \tag{2.49}$$

This is not the final form since we assumed isospin invariance, which we saw is equivalent to s-conservation if parity is conserved. Let us now demonstrate that s-conservation requires symmetry under exchange of $\sigma^{(1)}$ and $\sigma^{(2)}$, and, therefore, that the b-term cannot occur (for two neutrons or two protons this is true in general, even if isospin invariance were not satisfied). To prove the interdependence between symmetry under $\sigma^{(1)} \leftrightarrow \sigma^{(2)}$ exchange and s-conservation we consider the obvious relation between the $m_1 m_2$- and $s m_s$-representations of M:

$$\langle m_1' m_2' | M | m_1 m_2 \rangle = \sum_{ss'} C(\tfrac{1}{2}\tfrac{1}{2}s', m_1' m_2' m_s') C(\tfrac{1}{2}\tfrac{1}{2}s, m_1 m_2 m_s)$$
$$\times \langle s' m_s' | M | s m_s \rangle . \tag{2.50}$$

Now the interchange of $\sigma^{(1)}$ with $\sigma^{(2)}$ means the interchanges $m_1 \leftrightarrow m_2$ and $m_1' \leftrightarrow m_2'$, which can be handled through a symmetry relation of the Clebsch-Gordon coefficient:

$$C(j_1 j_2 j, m_1 m_2 m) = (-)^{j_1 + j_2 - j} C(j_2 j_1 j, m_2 m_1 m) . \tag{2.51}$$

Therefore the amplitude with $\sigma^{(1)}$ interchanged with $\sigma^{(2)}$ will be

$$\langle m_2' m_1' | M | m_2 m_1 \rangle = \sum_{ss'} C(\tfrac{1}{2}\tfrac{1}{2}s', m_1' m_2' m_s') C(\tfrac{1}{2}\tfrac{1}{2}s, m_1 m_2 m_s)(-)^{s+s'}$$
$$\times \langle s' m_s' | M | s m_s \rangle . \tag{2.52}$$

A sufficient condition for the equality of (2.50, 52) is the conservation of s. Incorporating isospin invariance, we finally end up with

$$M = a + c(\sigma^{(1)} + \sigma^{(2)})\hat{N} + m(\sigma^{(1)}\hat{N})(\sigma^{(2)}\hat{N}) + (g+h)(\sigma^{(1)}\hat{P})(\sigma^{(2)}\hat{P})$$
$$+ (g-h)(\sigma^{(1)}\hat{K})(\sigma^{(2)}\hat{K}) . \tag{2.53}$$

2.4 Examples for Spin Observables

Since the determination of the polarisation of a nucleon requires a special measurement, an experiment with an increasing total number of polarised particles in the initial and final state will be more and more difficult to perform. Therefore let us order the possible experiments with respect to this number. The simplest case is either an unpolarised initial state and the measurement of the polarisation of one of the outgoing particles or one polarised particle in the initial state (which will introduce a new direction in addition to the beam direction) and the measurement of the differential cross-section. Let us begin with the first case.

2.4.1 Polarisation

We assume the initial state to be unpolarised. That means

$$\langle S^\mu \rangle_i = 0 \tag{2.54}$$

for all S^μ besides $S^0 \equiv 1^{(1)} \otimes 1^{(2)}$. Therefore the density matrix in the initial state is just proportional to the unit matrix:

$$\varrho_i = \tfrac{1}{4} \mathrm{Tr}\{\varrho_i\} 1 . \tag{2.55}$$

The resulting spin averaged differential cross section I of (2.22) will be denoted by I_0:

$$I_0 = \tfrac{1}{4} \mathrm{Tr}\{MM^+\} . \tag{2.56}$$

The spin dependent nuclear force will lead to polarisations in the final state and we want to calculate the polarisation of one of the two particles:

$$P_0 \equiv \langle \sigma \rangle_f . \tag{2.57}$$

From the general expression (2.26) we get

$$I_0 P_0 = \tfrac{1}{4} \mathrm{Tr}\{MM^+ \sigma\} . \tag{2.58}$$

Since the particles are identical σ can be either $\sigma^{(1)}$ or $\sigma^{(2)}$. We can evaluate the right hand side using the Wolfenstein parametrisation (2.53) of M. In doing so we need several trace relations, which are easy to verify:

$$\mathrm{Tr}\{\sigma A\} = 0$$
$$\mathrm{Tr}\{\sigma A \, \sigma B\} = 4AB \tag{2.59}$$
$$\mathrm{Tr}\{\sigma A \, \sigma B \sigma C\} = 4\mathrm{i} A \times BC .$$

Excercise: Prove the trace-relations (2.59), where A, B and C are arbitrary, spin-independent vectors.

In this manner we can express the polarisation P_0 through the Wolfenstein parameters:

$$I_0 P_0 = \hat{N} 2 \mathrm{Re}\{c^*(a+m)\} . \tag{2.60}$$

Later in Sect. 2.5 we shall link the Wolfenstein parameters a, b, ... to scattering phase shifts in partial wave states, which can be calculated once the two-nucleon interaction V is given. Here we draw only the important conclusion that P_0 points in the normal direction of the scattering plane, which is the only spin-direction compatible with parity conservation.

2.4.2 Asymmetry

Let us now assume that either the beam or target particle is polarised in the initial state. This adds a further direction in space to the beam direction with respect to which one can measure the cross section. The initial density matrix is characterised now by the nonzero expectation value

$$P_i \equiv \langle \sigma \rangle_i \tag{2.61}$$

and according to (2.12) is given by

$$\varrho_i = \tfrac{1}{4} \text{Tr}\{\varrho_i\} \sum_{\varrho=0}^{3} \sigma_\varrho \langle \sigma_\varrho \rangle_i = \tfrac{1}{4} \text{Tr}\{\varrho_i\}(1 + \sigma P_i) . \tag{2.62}$$

Choosing $\mu = \nu = 0$ in the general expression (2.26), the cross section is

$$I = \tfrac{1}{4} \sum_\varrho \langle \sigma_\varrho \rangle_i \text{Tr}\{M \sigma_\varrho M^+\} = I_0 + \tfrac{1}{4} P_i \text{Tr}\{M \sigma M^+\} . \tag{2.63}$$

The trace can again be evaluated in much the same manner as in the previous case:

$$\tfrac{1}{4} \text{Tr}\{M \sigma M^+\} = 2 \hat{N} \text{Re}\{c^*(a + m)\} \tag{2.64}$$

and one finds

$$I = I_0 + P_i \hat{N} 2 \text{Re}\{c^*(a + m)\} . \tag{2.65}$$

We see that only the normal component of P_i leads to an additional nonzero contribution to the cross section. Again parity conservation rules out a dependence of I on components of P_i in the scattering plane. Assuming that P_i is transversal to the beam direction we encounter a situation as shown in Fig. 2.1.

The momenta q and q' define the scattering plane. If q' leaves to the left of the beam direction at the angle ϑ $\hat{N}$ points upwards, $\hat{N}_{(L)}$, whereas if q' leaves to the right at the same angle ϑ $\hat{N}$ points downwards, $\hat{N}_{(R)}$. Now having the direction P_i at our disposal, we can introduce an azimuthal angle φ which for spin-dependent forces is a dynamically relevant quantity. The φ-dependent part of I can be isolated by a left-right measurement:

$$A \equiv \frac{I(\vartheta, \varphi) - I(\vartheta, \varphi + \pi)}{I(\vartheta, \varphi) + I(\vartheta, \varphi + \pi)} = \frac{(P_i \hat{N}) 4 \text{Re}\{c^*(a + m)\}}{2 I_0} \tag{2.66}$$

or

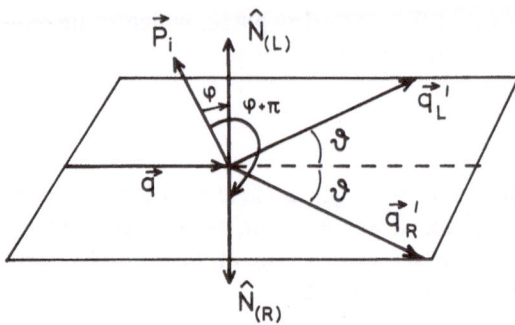

Fig. 2.1. The set-up for the left-right measurement of the asymmetry A

$$I_0 A = P_i \hat{N} 2 \mathrm{Re}\{c^*(a+m)\}. \tag{2.67}$$

Comparing with the expression (2.60) for the magnitude of the polarisation P_0 found in the previous section, we get

$$I_0 A = (P_i \hat{N}) I_0 P_0. \tag{2.68}$$

Specifically for 100% polarisation in the normal direction, $P_i \hat{N} = 1$, we have

$$A = P_0. \tag{2.69}$$

This equality is an important result [2.6, 7] and makes one of the two measurements superfluous. Even more interesting, it provides a means of searching for deviations from time reversal invariance which are linear in the violating amplitude. To demonstrate this, we keep the time reversal violating terms

$$t((\sigma^{(1)} P)(\sigma^{(2)} \hat{K}) + (\sigma^{(1)} \hat{K})(\sigma^{(2)} \hat{P})) \tag{2.70}$$

in the Wolfenstein parametrisation of M and work out $A - P_0$. This simple exercise shows that

$$A - P_0 \propto t. \tag{2.71}$$

This relation has been (and is still) or importance in testing time reversal invariance in the two-nucleon system [2.8].

2.4.3 Depolarisation

Having considered the two cases for one polarised particle, either in the initial or final state, we now regard the situation for two polarised particles. We begin with the situation where the polarisations of one particle each of the initial *and* final states are measured. In each case longitudinal or transversal polarisations can occur. This is conveniently described by using the unit vectors

$$q, \hat{N} \times q, \hat{N} \tag{2.72}$$

in the initial state and

$$\hat{P}, \hat{K}, \hat{N} \tag{2.73}$$

in the final state. In the lab-system the three vectors in the final state have a simple meaning, as we shall show now. The individual particle momenta k_1, k_2 and k_1', k_2' in the initial and final state, respectively, are related to the total momentum K and the relative momenta q and q' through

$$
\begin{aligned}
k_1 &= K/2 + q, & k_1' &= K/2 + q' \\
k_2 &= K/2 - q, & k_2' &= K/2 - q'.
\end{aligned}
\tag{2.74}
$$

Specifically in the lab-system, defined by $k_2 = 0$, we have

$$
\begin{aligned}
q &= \tfrac{1}{2}k_1 \\
q' &= \tfrac{1}{2}(k_1' - k_2') = k_1' - \tfrac{1}{2}k_1,
\end{aligned}
\tag{2.75}
$$

which leads to

$$k_1' = q + q' \quad \text{or} \quad \hat{k}_1' = \hat{P}. \tag{2.76}$$

Thus $\hat{P}$ points into the direction of the final momentum of particle 1 and can serve to define the longitudinal polarisation of particle 1. The scattering plane is of course common to the center-of-mass and lab-system, and therefore

$$\hat{N} = \hat{k}_1 \times \hat{k}_1' = \hat{q} \times \hat{q}' \tag{2.77}$$

serves for both systems. Finally the transversal direction in the scattering plane is

$$\hat{N} \times \hat{k}_1' = \hat{N} \times \hat{P} = \hat{K}, \tag{2.78}$$

which together with $\hat{N}$ completely describes an arbitrary transversal polarisation.

It remains to establish the connection between the two sets of units vectors (2.72, 73):

$$
\begin{aligned}
\hat{N} \times \hat{q} &= \cos\frac{\vartheta}{2}\,\hat{K} + \sin\frac{\vartheta}{2}\,\hat{P} \\
\hat{q} &= -\sin\frac{\vartheta}{2}\,\hat{K} + \cos\frac{\vartheta}{2}\,P.
\end{aligned}
\tag{2.79}
$$

This is left as an exercise.

Exercise: Prove (2.79) together with the relation between the lab- and center-of-mass scattering angle:

$$\cos \frac{\vartheta}{2} = \cos \vartheta_{lab} \,.$$

With the kinematical preliminaries out of the way, it is now a straightforward exercise to calculate the polarisation of one final particle arising from an initial state where one particle is polarised. According to (2.26) we get

$$I \langle \sigma \rangle_f = \tfrac{1}{4} \sum_\lambda \langle \sigma_\lambda \rangle_i \text{Tr} \{ M \sigma_\lambda M^+ \sigma \} \tag{2.80}$$

or

$$I \langle \sigma \rangle_f = \tfrac{1}{4} \text{Tr} \{ M M^+ \sigma \} + \tfrac{1}{4} \text{Tr} \{ M (P_i \sigma) M^+ \sigma \} \,. \tag{2.81}$$

In the first term we recognize

$$I_0 P_0 \hat{N} = \tfrac{1}{4} \text{Tr} \{ M M^+ \sigma \} \,. \tag{2.82}$$

The second term due to P_i will be expanded completely by writing

$$P_i = \hat{q} (\hat{q} P_i) + \hat{N} \times \hat{q} (\hat{N} \times \hat{q} P_i) + \hat{N} (\hat{N} P_i) \tag{2.83}$$

and

$$\sigma = \hat{N} (\hat{N} \sigma) + \hat{K} (\hat{K} \sigma) + \hat{P} (\hat{P} \sigma) \,. \tag{2.84}$$

Then the trace in (2.81) decomposes into

$$\begin{aligned}
\text{Tr} & \{ M (P_i \sigma) M^+ \sigma \} \\
& = (\hat{q} P_i) [N \underline{\text{Tr} \{ M (\hat{q} \sigma) M^+ (\hat{N} \sigma) \}} \\
& \quad + \hat{K} \text{Tr} \{ M (\hat{q} \sigma) M^+ (\hat{K} \sigma) \} + \hat{P} \text{Tr} \{ M (\hat{q} \sigma) M^+ (\hat{P} \sigma) \}] \\
& \quad + (\hat{N} \times \hat{q} P_i) [\hat{N} \underline{\text{Tr} \{ M (\hat{N} \times \hat{q} \sigma) M^+ (\hat{N} \sigma) \}} \\
& \quad + \hat{K} \text{Tr} \{ M (\hat{N} \times q \sigma) M^+ (\hat{K} \sigma) \} + \hat{P} \text{Tr} \{ M (\hat{N} \times \hat{q} \sigma) M^+ (\hat{P} \sigma) \}] \\
& \quad + (\hat{N} P_i) [\hat{N} \text{Tr} \{ M (\hat{N} \sigma) M^+ (\sigma \hat{N}) \} \\
& \quad + \hat{K} \underline{\text{Tr} \{ M (\hat{N} \sigma) M^+ (\sigma \hat{K}) \}} + \hat{P} \underline{\text{Tr} \{ M (\hat{N} \sigma) M^+ (\sigma \hat{P}) \}}] \,.
\end{aligned} \tag{2.85}$$

Among the various terms, the ones underlined are zero because of parity conservation. Since that invariance is built into M, this is of course an automatic result in calculating the traces. The remaining ones carry the following standard notation [2.9]:

$$I_0 D \equiv \tfrac{1}{4} \mathrm{Tr}\{M(\sigma \hat{N}) M^+(\sigma \hat{N})\}$$
$$I_0 R \equiv \tfrac{1}{4} \mathrm{Tr}\{M(\sigma \hat{N} \times \hat{q}) M^+(\sigma \hat{K})\}$$
$$I_0 R' \equiv \tfrac{1}{4} \mathrm{Tr}\{M(\sigma \hat{N} \times \hat{q}) M^+(\sigma \hat{P})\} \tag{2.86}$$
$$I_0 A \equiv \tfrac{1}{4} \mathrm{Tr}\{M(\sigma \hat{q}) M^+(\sigma \hat{K})\}$$
$$I_0 A' \equiv \tfrac{1}{4} \mathrm{Tr}\{M(\sigma \hat{q}) M^+(\sigma \hat{P})\} \, .$$

The polarisation P is then given as

$$I_0 P = I_0 \{\hat{N}[P_0 + D(\hat{N} P_i)] + \hat{P}[A'(\hat{q} P_i) + R'(\hat{N} \times \hat{q} P_i)]$$
$$+ \hat{K}[A(\hat{q} P_i) + R(N \times \hat{q} P_i)]\} \, . \tag{2.87}$$

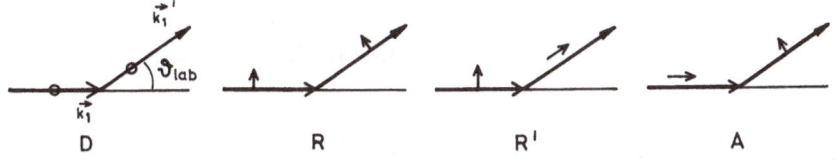

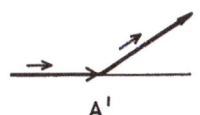

Fig. 2.2. The definition of the depolarisation parameters D, R, R', A, and A'

The content of (2.87) is depicted in Fig. 2.2. Whereas R, R', A, and A' describe the four possible spin-nonflip and spin-flip combinations in the scattering plane, the quantity D determines the change in polarisation in the normal direction. Again we recognize that an initial polarisation P_i in the normal direction cannot flip into the scattering plane and vice versa. Finally we have to evaluate the traces in (2.86), and one finds [2.1]:

$$I_0 D = |a|^2 + |m|^2 + 2|c|^2 - 2|g|^2 - 2|h|^2$$
$$I_0 R' = \sin \tfrac{\vartheta}{2}(|a|^2 - |m|^2 - |g-h|^2 + |g+h|^2) + \cos \tfrac{\vartheta}{2} \mathrm{Im}\{2c(a^* - m^*)\}$$
$$I_0 R = \cos \tfrac{\vartheta}{2}(|a|^2 - |m|^2 + |g-h|^2 - |g+h|^2) - \sin \tfrac{\vartheta}{2} \mathrm{Im}\{2c(a^* - m^*)\} \tag{2.88}$$
$$I_0 A = -\sin \tfrac{\vartheta}{2}(|a|^2 - |m|^2 + |g-h|^2 - |g+h|^2) - \cos \tfrac{\vartheta}{2} \mathrm{Im}\{2c(a^* - m^*)\}$$
$$I_0 A' = \cos \tfrac{\vartheta}{2}(|a|^2 - |m|^2 + |g+h|^2 - |g-h|^2) - \sin \tfrac{\vartheta}{2} \mathrm{Im}\{2c(a^* - m^*)\} \, .$$

2.4.4 Spin Correlation Parameters

Staying with the case of two polarised particles, we choose them now to be measured in coincidence in the final state. In the lab-system Fig. 2.3 shows the possibilities allowed by parity invariance.

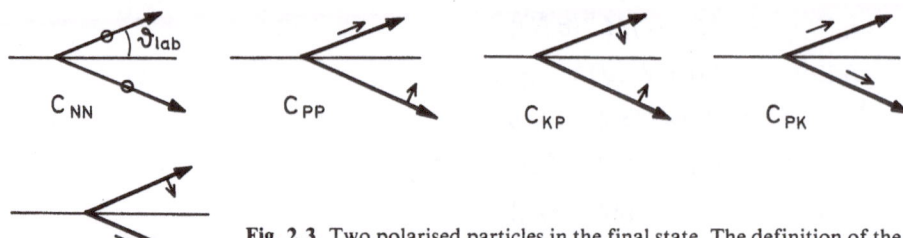

Fig. 2.3. Two polarised particles in the final state. The definition of the various spin correlation coefficients C

The analytic expression for the two polarisations to occur in coincidence, starting from an unpolarised initial state, is given by

$$I\langle\sigma^{(1)}\sigma^{(2)}\rangle_f = \tfrac{1}{4}\text{Tr}\{MM^+\,\sigma^{(1)}\sigma^{(2)}\}. \tag{2.89}$$

Again we decompose $\sigma^{(1)}$ and $\sigma^{(2)}$ into the final triple of directions and using parity conservation we get

$$I\langle\sigma^{(1)}\sigma^{(2)}\rangle_f = I_0(C_{NN}\hat{N}\hat{N} + C_{PP}\hat{P}\hat{P} + C_{PK}\hat{P}\hat{K} + C_{KP}\hat{K}\hat{P} + C_{KK}\hat{K}\hat{K}), \tag{2.90}$$

where for instance

$$\begin{aligned} I_0 C_{NN} &= \tfrac{1}{4}\text{Tr}\{MM^+\,\sigma^{(1)}\hat{N}\sigma^{(2)}\hat{N}\} \\ I_0 C_{PK} &= \tfrac{1}{4}\text{Tr}\{MM^+\,\sigma^{(1)}\hat{P}\sigma^{(2)}\hat{K}\} \end{aligned} \tag{2.91}$$

etc.

Exercise: Determine the C-coefficients in terms of the Wolfenstein parameters.

It remains to consider the case that both the beam and target particles are polarised in the initial state. Then according to (2.26) the scattering cross section will be

$$I = I_0 + \tfrac{1}{4}\sum_{k,l}\langle\sigma_k^{(1)}\sigma_l^{(2)}\rangle_i\,\text{Tr}\{M\sigma_k^{(1)}\sigma_l^{(2)}M^+\}. \tag{2.92}$$

By now the recipe should be clear. One decomposes $\sigma^{(1)}$ and $\sigma^{(2)}$ into the 3 initial directions (2.72) which will lead to spin correlation parameters A_{xy} expressed by certain traces, where $\hat{x}$, $\hat{y}$ can point in any of the three initial directions. To separate the individual A_{xy} from I_0 and each other linear combinations of cross sections in certain directions have to be taken. Further the groups of A_{xy}- and C-coefficients are again related by time reversal invariance [2.10] and deviations thereof can be used to measure violations of that invariance.

It is a straightforward exercise to introduce higher order spin correlation coefficients by treating the case of a total number of three and four polarised particles. One can expect that they explore the spin dependence of the nuclear force in even greater detail.

2.5 Partial-Wave Decomposition

We now have to face the actual calculation of the Wolfenstein-parameters, once the two-nucleon interaction V is given. Since V is rotationally invariant, the scattering states can be classified into states of fixed total angular momentum, and consequently the M-matrix will decompose into parts which each belong to a fixed total angular momentum. We shall see that each of these partial-wave M-matrix elements can be characterized by a few phase shift parameters. Moreover at low energies up to a few hundred MeV, the two nucleons only feel the nuclear interaction — which is very short range — in a relatively small number of angular momentum states, while in higher angular momentum states they pass each other without interaction. Therefore the partial-wave decomposition is an advantageous parametrisation of the M-matrix and results in a small number of terms.

The strategy will be now the following one. First we invert the Wolfenstein parametrisation (2.53) and express the parameters a, b, ... in terms of the M-matrix elements. The M-matrix elements are taken between momentum and spin-eigenstates, which can be decomposed into states of good angular momentum. This leads then to the desired partial-wave representation of M. On the way we shall also encounter the partial-wave representation of the Lippmann-Schwinger equation, which will be the dynamical equation one must finally solve.

The inversion of (2.53) is easily achieved, since the individual terms are orthogonal with respect to the trace. Thus, as a first step, we get

$$a = \tfrac{1}{4}\text{Tr}\{M\}$$

$$c = \tfrac{1}{8}\text{Tr}\{M(\sigma^{(1)}+\sigma^{(2)})\hat{N}\}$$

$$m = \tfrac{1}{4}\text{Tr}\{M(\sigma^{(1)}\hat{N})(\sigma^{(2)}\hat{N})\} \qquad (2.93)$$

$$g+h = \tfrac{1}{4}\text{Tr}\{M\sigma^{(1)}\hat{P}\sigma^{(2)}\hat{P}\}$$

$$g-h = \tfrac{1}{4}\text{Tr}\{M\sigma^{(1)}\hat{K}\sigma^{(2)}\hat{K}\}.$$

Let us choose a specific coordinate system such that the $x-z$ plane coincides with the scattering plane as depicted in Fig. 2.4.

Then the various momenta are given as

$$\hat{q} = \begin{pmatrix} 0 \\ 0 \\ 1 \end{pmatrix} \quad \hat{q}' = \begin{pmatrix} \sin\vartheta \\ 0 \\ \cos\vartheta \end{pmatrix} \quad \hat{N} = \begin{pmatrix} 0 \\ 1 \\ 0 \end{pmatrix} \quad \hat{K} = \begin{pmatrix} \cos\vartheta/2 \\ 0 \\ -\sin\vartheta/2 \end{pmatrix} \quad \hat{P} = \begin{pmatrix} \sin\vartheta/2 \\ 0 \\ \cos\vartheta/2 \end{pmatrix} .$$

$$(2.94)$$

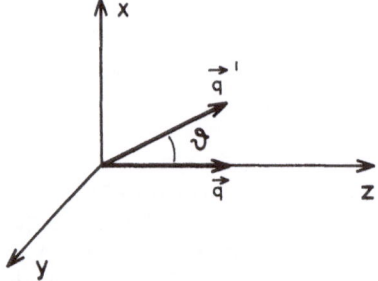

Fig. 2.4. The initial and final relative momenta, q and q', respectively, spanning the x-z-plane

Since s is conserved, a further useful step is to rewrite the M-matrix from the $m_1 m_2$-representation into the $s m_s$-representation, $M^s_{m'_s m_s}$. This is easily accomplished using (2.50) and one gets explicitly

$$(M_{m'_1 m'_2 m_1 m_2}) = \begin{pmatrix} M^1_{11} & \frac{1}{\sqrt{2}}M^1_{10} & \frac{1}{\sqrt{2}}M^1_{10} & M^1_{1-1} \\ \frac{1}{\sqrt{2}}M^1_{01} & \frac{1}{2}(M^0_{00}+M^1_{00}) & \frac{1}{2}(-M^0_{00}+M^1_{00}) & \frac{1}{\sqrt{2}}M^1_{0-1} \\ \frac{1}{\sqrt{2}}M^1_{01} & \frac{1}{2}(-M^0_{00}+M^1_{00}) & \frac{1}{2}(M^0_{00}+M^1_{00}) & \frac{1}{\sqrt{2}}M^1_{0-1} \\ M^1_{-11} & \frac{1}{\sqrt{2}}M^1_{-10} & \frac{1}{\sqrt{2}}M^1_{-10} & M^1_{-1-1} \end{pmatrix}.$$

$$(2.95)$$

Inserting now the Pauli spin matrices, the explicit form for the various momenta and (2.95) into (2.93), it is a straightforward exercise to arrive at

$$a = \tfrac{1}{4}(M^1_{11}+M^1_{-1-1}+M^0_{00}+M^1_{00})$$

$$c = \frac{i}{4\sqrt{2}}(M^1_{10}-M^1_{01}+M^1_{0-1}-M^1_{-10})$$

$$m = \tfrac{1}{4}(-M^1_{1-1}-M^1_{-11}+M^0_{00}-M^1_{00})$$

$$g = \tfrac{1}{8}(M^1_{11}+M^1_{1-1}+M^1_{-11}+M^1_{-1-1}-2M^0_{00})$$

$$h = \tfrac{1}{8}\cos\vartheta(M^1_{11}+M^1_{1-1}-M^1_{-11}+M^1_{-1-1}-2M^0_{00})$$

$$+\frac{1}{8}\frac{\sin\vartheta}{\sqrt{2}}(2M^1_{10}+2M^1_{01}-2M^1_{0-1}-2M^1_{-10}) .$$

$$(2.96)$$

The main step is now the decomposition of $M^s_{m'_s m_s}$ given by

$$M^s_{m'_s m_s} = -\mu(2\pi)^2 \langle q' | \langle s m_{s'} | \langle t m_t | V | q s m_{s_t} t m_t \rangle_a^{(+)} \tag{2.97}$$

into partial-wave states. The states of conserved total angular momentum are built up by orbital motion and spin as

$$\mathscr{Y}^{JM}_{ls}(\hat{x}) = \sum_{m_l m_s} C(ls J, m_l m_s M) \, Y_{lm_l}(\hat{x}) \, \chi_{sm_s}. \tag{2.98}$$

Because of the tensor force in the nuclear interaction, the orbital angular momentum l itself is not conserved. The decomposition is easily performed for the free states recalling (1.162) and using (2.98)

$$\psi^0_q(x)\chi_{sm_s} = \frac{4\pi}{(2\pi)^{3/2}} \sum_{lm_l} Y_{lm_l}(\hat{x}) i^l j_l(qx) Y^*_{lm_l}(\hat{q}) \chi_{sm_s}$$

$$= \frac{4\pi}{(2\pi)^{3/2}} \sum_{lJM} C(ls J, M-m_s, m_s, M) i^l j_l(qx) Y^*_{lM-m_s}(\hat{q}) \, \mathscr{Y}^{JM}_{ls}(\hat{x}). \tag{2.99}$$

The decomposition of the scattering state in (2.97) needs a more detailed discussion, since at the same time we want to achieve the partial-wave representation of the Lippmann-Schwinger equation. It is well known [2.11] that the free Green's function $G_0(x, x', E)$ can be decomposed as

$$G_0(x, x', E) = \sum_{lm_l} Y_{lm_l}(\hat{x}) G_l(x, x', E) Y^*_{lm_l}(\hat{x}'), \tag{2.100}$$

where the radial Green's function is given as

$$G_l(x, x', E) = -2\mu i \sqrt{2\mu E} \, j_l(qr_<) h_l^{(1)}(qr_>) \tag{2.101}$$

and j_l and $h_l^{(1)}$ are spherical Bessel functions [2.12]:

$$j_l(z) = \sqrt{\frac{\pi}{2z}} \, J_{l+1/2}(z)$$

$$h_l^{(1)}(z) = \sqrt{\frac{\pi}{2z}} \, H^{(1)}_{l+1/2}(z). \tag{2.102}$$

We may insert a complete set of states in spin space into (2.100) and rewrite it as

$$G_0(x, x', E) = \sum_{JlsM} \mathscr{Y}^{JM}_{ls}(\hat{x}) G_l(x, x', E) \, \mathscr{Y}^{JM*}_{ls}(\hat{x}'). \tag{2.103}$$

The final step is to decompose the scattering state as well

$$\Psi^{(+)}_{qsm_s}(x) = \sum_{lJM} \mathscr{Y}^{JM}_{ls}(\hat{x}) \, \Psi^{JM}_{ls}(x). \tag{2.104}$$

Inserting now (2.99), (2.103) and (2.104) into the Lippmann-Schwinger equation and projecting onto the states $\mathscr{Y}_{ls}^{JM}$ we arrive immediately at

$$\Psi_{ls}^{JM}(x) = \frac{4\pi}{(2\pi)^{3/2}} C(lsJ, M-m_s, m_s, M) i^l Y_{lM-m_s}^*(\hat{q}) j_l(qx)$$

$$+ \int_0^\infty dx' x'^2 G_l(xx'E) \sum_{l'} (\mathscr{Y}_{ls}^{JM}|V|\mathscr{Y}_{l's}^{JM}) \Psi_{l's}^{JM}(x') . \qquad (2.105)$$

This is a coupled system of integral equations for the radial scattering wave functions $\Psi_{ls}^{JM}(x)$. Now each orbital angular momentum can initiate a scattering process. Therefore let us introduce *new* radial wave functions, which by definition have an incoming wave in the state l only. The nuclear force will then produce scattered parts in the same l as well as certain other l's. Therefore the new amplitudes $\Psi_{l's, ls}^J$ will have an additional index pair, ls, indicating the state of the incoming wave. They are defined through

$$\Psi_{l's, ls}^J(x) = \delta_{ll'} j_l(qx) + \int_0^\infty dx' x'^2 G_{l'}(xx'E) \sum_{l''} (\mathscr{Y}_{l's}^{JM}|V|\mathscr{Y}_{l''s}^{JM}) \Psi_{l''s, ls}^J(x') . \qquad (2.106)$$

Obviously this notation would immediately allow for the generalisation to transitions in the total spin s as well.

How are the radial wave functions $\Psi_{ls}^{JM}(x)$ for the scattering state obeying the set (2.105) related to the auxiliary ones defined in (2.106)? We multiply both sides of (2.106) by

$$\frac{4\pi}{(2\pi)^{3/2}} C(lsJ, M-m_s, m_s, M) i^l Y_{lM-m_s}^*(\hat{q})$$

and sum over l, then obviously the set (2.105) results, now for the radial wave function

$$\sum_l \frac{4\pi}{(2\pi)^{3/2}} C(lsJ, M-m_s m_s M) i^l Y_{lM-m_s}^*(\hat{q}) \Psi_{l's, ls}^J(x) .$$

Since however (2.105) has a unique solution, the desired connection is

$$\Psi_{l's}^{JM}(x) = \sum_l \frac{4\pi}{(2\pi)^{3/2}} C(lsJ, M-m_s m_s M) i^l Y_{lM-m_s}^*(\hat{q}) \Psi_{l's\,ls}^J(x) . \qquad (2.107)$$

Combining (2.104) and (2.107), the antisymmetric scattering state including a state of fixed isospin t is

$$|qsm_stm_t\rangle_a^{(+)} \equiv (1-P_{12})|qsm_stm_t\rangle^{(+)}$$

$$= \frac{4\pi}{(2\pi)^{3/2}} \sum_{Jl'Ml} |\mathcal{Y}_{l's}^{JM}\rangle |tm_t\rangle |\Psi_{l'sls}^J\rangle$$

$$\times C(lsJ, M-m_sm_sM) i^l Y_{lM-m_s}^*(\hat{q})[1-(-)^{l+s+t}] . \quad (2.108)$$

The antisymmetriser $(1-P_{12})$ becomes $[1-(-)^{l+s+t}]$ for each partial wave state characterised by the quantum numbers lst. Therefore, only states for which $l+s+t$ is odd are allowed by the Pauli principle.

Let us now elaborate the information contained in the asymptotic form of the radial scattering wave functions. This is sufficient to build up the expressions for the observables. In order to evaluate (2.105) for $x \to \infty$ we note that

$$\left.\begin{array}{l} j_l(z) \rightarrow \dfrac{\sin(z-\frac{1}{2}l\pi)}{z} \quad \text{and} \\[2em] h_l^{(1)}(z) \rightarrow \dfrac{\exp[i(z-\frac{1}{2}l\pi)]}{z} \end{array}\right\} \quad \text{for} \quad z \to \infty . \quad (2.109)$$

Therefore the auxiliary, radial, scattering wave functions behave asymptotically as

$$\Psi_{l'sls}^J(x) \rightarrow \delta_{ll'} \frac{\sin(qx-\frac{1}{2}l\pi)}{qx}$$

$$-2\mu q \frac{\exp[i(qx-\frac{1}{2}l\pi)]}{qx} \sum_{l''} \langle \mathcal{Y}_{l's}^{JM} j_{l'} |V| \mathcal{Y}_{l''s}^{JM} \Psi_{l''sls}^J\rangle$$

$$\equiv -\frac{1}{2iqx}(\exp[-i(qx-\frac{1}{2}l\pi)]\delta_{ll'} - \exp[i(qx-\frac{1}{2}l'\pi)]S_{l'sls}^J) . \quad (2.110)$$

This is a superposition of a radially incoming wave in state l and radially outgoing waves in all states l'. The amplitude of the outgoing wave can be read off from (2.110) to be

$$S_{l'sls}^J = \delta_{ll'} - 4i\mu q \sum_{l''} \langle \mathcal{Y}_{l's}^{JM} j_{l'} |V| \mathcal{Y}_{l''s}^{JM} \Psi_{l''sls}^J\rangle \quad (2.111)$$

and is called the partial wave S-matrix element. We shall see below that this notation is indeed justified.

We are now fully prepared to write down the partial wave decomposition of the M-matrix. Inserting (2.99) and (2.108) into (2.97) we get

$$M^s_{m'_s m_s} = -\mu(2\pi)^2 \sum_{Jl'M} \frac{4\pi}{(2\pi)^{3/2}} C(l'sJ, M - m_s m_s M)$$

$$\times i^{-l'} Y_{l'M-m_s}(\hat{q}') \sum_{l''l} \langle \mathcal{Y}^{JM}_{l's} j_{l'} | \langle t m_t | V | \mathcal{Y}^{JM}_{l''s} \Psi^J_{l''sls} \rangle | t m_t \rangle$$

$$\times C(lsJ, M - m_s m_s M) i^l Y^*_{lM-m_s}(\hat{q}) [1 - (-)^{l+s+t}] \frac{4\pi}{(2\pi)^{3/2}} \quad (2.112)$$

or in terms of the S-matrix elements:

$$M^s_{m'_s m_s} = \frac{2\pi}{iq} \sum_{Jl'lM} C(l'sJ, M - m'_s m'_s M) Y_{l'M-m_s}(\hat{q}')$$

$$\times i^{-l'+l}(S^J_{l'sls} - \delta_{l'l}) C(lsJ, M - m_s m_s M) Y^*_{lM-m_s}(\hat{q}) [1 - (-)^{l+s+t}]. \quad (2.113)$$

Since we choose the z-axis to be in the $\hat{q}$-direction,

$$Y^*_{l,M-m_s}(\hat{q}) = \delta_{Mm_s} \sqrt{\frac{2l+1}{4\pi}} \quad (2.114)$$

and we end up with

$$M^s_{m'_s m_s} = \frac{1}{iq} \sum_{Jll'} C(l'sJ, m_s - m'_s m'_s m_s) Y_{l'm'_s-m_s}(\hat{q}')$$

$$\times i^{-l'+l}(S^J_{l'sls} - \delta_{l'l}) C(lsJ, 0 m_s) \sqrt{\pi(2l+1)}\, [1 - (-)^{l+s+t}]. \quad (2.115)$$

This relation is the important link between M-matrix elements, which determine directly the Wolfenstein parameters and therefore the observables on the one side, and the S-matrix elements in partial wave states on the other side. It will be the task of the remaining sections to establish the properties of S and to present techniques to calculate them for a given nuclear interaction V.

Since we choose q' to lie in the x-z plane (Fig. 2.4), the azimuthal angle φ of q' is zero and the spherical harmonics in (2.115) are real. It is then a simple exercise to verify the properties

$$M^1_{0-1} = -M^1_{01}$$
$$M^1_{1-1} = M^1_{-11}$$
$$M^1_{-10} = -M^1_{10}$$
$$M^1_{11} = M^1_{-1-1}.$$

$$(2.116)$$

This yields a slight simplification of the relations (2.96).

In (2.115) again the selective factor for the partial wave states allowed by the Pauli principle is present. Since M refers to a fixed isospin t it is already the physical amplitude for the (pp) and (nn) systems, which are pure $t = 1$ states. The sums over l and l' vary therefore only over those values such that $(l+s) = $ even. In the case of (pn), both $t = 0$ and $t = 1$ occur, and the physical amplitude is $1/2$ the sum of $M^{t=0}$ and $M^{t=1}$ as we saw in (2.29). Therefore the bracket $(1 - (-)^{l+s+t})$ cancels against the factor $1/2$ and one sums (2.115) unrestricted over all l and l', to get the physical (np) amplitude.

2.6 Standard S-Matrix Representations

We have established the following chain

$$S^J_{l'sls} \rightarrow M^s_{m'_s m_s} \rightarrow a, b, \ldots \rightarrow \text{observables} .$$ (2.117)

Since the full dynamical information is contained in the S-matrix elements $S^J_{l'sls}$, the general properties of $S^J_{l'sls}$ are of great importance. Let us first justify the notation partial-wave S-matrix element. The S-matrix elements with respect to momentum eigenstates were defined in Sect. 1.5. Adding the spin- and isospin space, this relation is generalized to

$$\hat{S}^t_{q'sm'_st, \, qsm_st} = \delta_{m'_s m_s}[\delta(\hat{q}' - \hat{q}) - (-)^{s+t}\delta(\hat{q}' + \hat{q})] - 2i\pi\mu q \, T_{q'sm'_st, \, qsm_st}$$ (2.118)

or using the connection (2.34) between the T- and M-matrix-elements it is

$$\hat{S}_{q'sm'_st, \, qsm_st} = \delta_{m'_s m_s}[\delta(\hat{q}' - \hat{q}) - (-)^{s+t}\delta(\hat{q}' + \hat{q})] + \frac{iq}{2\pi} M^{st}_{m'_s m_s}(q', q) .$$ (2.119)

The free term is clearly $\langle q'sm'_s t m_t | q s m_s t m_t \rangle_a$ divided by $\delta(E_q - E_q)/mq$.

If our expectation is correct, the terms proportional to $\delta_{l'l}$ in the decomposition (2.113) have to cancel the first term on the right hand side of (2.119). That part of M is

$$-\frac{2\pi}{iq} \sum_{JlM} C(lsJ, M - m'_s m_s M) \, Y_{l,M-m'_s}(\hat{q}')$$

$$\times C(lsJ, M - m_s m_s M) \, Y^*_{l,M-m_s}(\hat{q})[1 - (-)^{l+s+t}]$$

$$= -\frac{2\pi}{iq} \sum_{lM} \delta_{m'_s m_s} Y_{l,M-m_s}(\hat{q}') \, Y^*_{l,M-m_s}(\hat{q})[1 - (-)^{l+s+t}]$$

$$= -\frac{2\pi}{iq} \delta_{m'_s m_s} \sum_{lM} [Y_{lM}(q') \, Y^*_{lM}(\hat{q}) - (-)^{s+t} Y_{lM}(\hat{q}') \, Y^*_{lM}(-\hat{q})]$$

$$= -\frac{2\pi}{iq} \delta_{m'_s m_s}[\delta(\hat{q}' - \hat{q}) - (-)^{s+t}\delta(\hat{q}' + \hat{q})] .$$ (2.120)

In the first equality we used the orthogonality relation of the Clebsch-Gordon coefficients

$$\sum_J C(lsJ, M-m'_s m'_s M) C(lsJ, M-m_s m_s M) = \delta_{m'_s m_s} \tag{2.121}$$

and in the third equality the completeness relation of the spherical harmonics. Multiplying by $iq/2\pi$ as required by (2.119) this expression indeed cancels the potential free term in (2.119) and we get the partial-wave decomposition of the S-matrix element:

$$\hat{S}_{q'sm'_s l, qsm_s l} = \sum_{Jl'lM} C(l'sJ, M-m'_s m'_s M) Y_{l'M-m'_s}(\hat{q}') i^{-l'+l}$$

$$\times S^J_{l'sls} C(lsJ, M-m_s m_s M) Y^*_{lM-m_s}(\hat{q}) [1-(-)^{l+s+t}] . \tag{2.122}$$

Let us first exploit the consequences of the assumption of time-reversal invariance as expressed in (2.43). This reads in the sm_s-representation

$$(-)^{-m_s-m'_s} M^s_{-m_s-m'_s}(-q, -q') = M^s_{m'_s m_s}(q', q) , \tag{2.123}$$

which according to (2.119) carries over immediately to

$$(-)^{-m_s-m'_s} \hat{S}_{-qs-m_s l, -q's-m'_s l} = \hat{S}_{q'sm'_s l, qsm_s l} . \tag{2.124}$$

Inserting the decomposition (2.122) into both sides of (2.124) it is an easy exercise, using the orthogonality of the spherical harmonics and of the Clebsch-Gordon coefficients, to reduce (2.124) to

$$S^J_{l'sls} = S^J_{lsl's} . \tag{2.125}$$

Thus for a time-reversal invariant potential V, the partial-wave S-matrix elements can be chosen to be symmetric.

Exercise: Verify (2.125).

The second property of $S_{l'sls}$ is unitarity:

$$\sum_{l''} S^{J*}_{l''sl's} S^J_{l''sls} = \delta_{l'l} . \tag{2.126}$$

This relation is of course just the partial wave representation of the unitarity relation established in Sect. 1.5 and generalized in an obvious manner including spin and isospin to

$$\frac{1}{2} \int \sum_{m_s} d\hat{q}'' \hat{S}^*_{\hat{q}''sm''_s l, \hat{q}'sm'_s l} \hat{S}_{\hat{q}''sm''_s l, \hat{q}sm_s l} = \delta_{m'_s m_s}(\delta(\hat{q}'-\hat{q})-(-)^{s+t}\delta(\hat{q}'+\hat{q})) . \tag{2.127}$$

The factor 1/2 on the left side accounts for the missing normalisation factor in the definition of the antisymmetrised states (2.30). Indeed, assume that the interaction is turned off. Then the $\hat{S}$-matrix element will be just

$$S^0_{\hat{q}''sm_s''t,\, \hat{q}sm_st} = \langle \hat{q}''sm_s''t | \hat{q}sm_st \rangle_a \tag{2.128}$$

and (2.127) would turn into

$$\tfrac{1}{2}\int d\hat{q}'' \sum_{m_s''} {}_a\langle \hat{q}'sm_s't | \hat{q}''sm_s''t\rangle\langle \hat{q}''sm_s''t | \hat{q}sm_st\rangle_a$$
$$= \tfrac{1}{2}{}_a\langle \hat{q}'sm_s't | \hat{q}sm_st\rangle_a = \langle \hat{q}'sm_s't | \hat{q}sm_st\rangle_a$$
$$= \delta_{m_s'm_s}[\delta(\hat{q}-\hat{q}')-(-)^{s+t}\delta(\hat{q}+\hat{q}')] \,. \tag{2.129}$$

Exercise: Derive (2.126) from (2.127).

It is a worthwhile exercise to verify the unitarity relation (2.126) directly from the radial equations (2.106). Let us first turn that coupled set of integral equations into coupled differential equations. Consider the radial part of the kinetic energy

$$T_l(x) \equiv -\frac{1}{2\mu}\frac{1}{x^2}\frac{d}{dx}\left(x^2\frac{d}{dx}\right)+\frac{1}{2\mu}\frac{l(l+1)}{x^2}\,. \tag{2.130}$$

Since $G_0(x, x', E)$ obeys (1.78), its radial part $G_l(x, x', E)$ fulfils

$$(T_l(x)-E)\,G_l(x, x', E) = -\,\delta(x-x')/xx'\,. \tag{2.131}$$

Let us apply $[T_l(x)-E]$ to (2.106). The result is a coupled set of differential equations

$$[T_{l'}(x)-E]\,\Psi^J_{l'sls}(x)+ \sum_{l''} \mathscr{V}^J_{l'sl''s}(x)\,\Psi^J_{l''sls}(x) = 0 \tag{2.132}$$

with

$$\mathscr{V}^J_{l'sls}(x) \equiv \langle \mathscr{Y}^{JM}_{l's} | V | \mathscr{Y}^{JM}_{ls}\rangle\,. \tag{2.133}$$

Now we can follow the standard procedure of establishing flux conservation. We multiply (2.132) from the left by $\Psi^{J*}_{l'sl'''s}(x)$, sum over l' and integrate between 0 and R. Then we subtract the conjugate complex with l and l''' interchanged. We obtain

$$\sum_{l'}\int_0^R x^2\,dx\,\{\Psi^{J*}_{l'sl'''s}(x)\,[T_{l'}(x)-E]\,\Psi^J_{l'sls}(x) - \Psi^J_{l'sls}(x)\,[T_{l'}(x)-E]\,\Psi^{J*}_{l'sl'''s}(x)\}$$

$$+ \sum_{l'l''}\int_0^R x^2\,dx\,\Psi^{J*}_{l'sl'''s}(x)\,\mathscr{V}^J_{l'sl''s}(x)\,\Psi^J_{l''sls}(x)$$

$$- \sum_{l'l''}\int_0^R x^2\,dx\,\Psi^J_{l'sls}(x)\,\mathscr{V}^J_{l'sl''s}(x)\,\Psi^{J*}_{l''sl'''s}(x) = 0\,. \tag{2.134}$$

In the limit $R\to\infty$ the last two terms cancel, since $\mathscr{V}$ is assumed to be hermitean. The differentiations in $T_l(x)$ simplify through the usual trick of putting $\Psi(x) \equiv u(x)/x$, and we are left with

$$\lim_{R\to\infty} \sum_{l'} \int_0^R x^2 dx \left[u^{J*}_{l'sl''s}(x) \frac{d^2}{dx^2} u^J_{l'sls}(x) - u^J_{l'sls}(x) \frac{d^2}{dx^2} u^{J*}_{l'sl''s}(x) \right] = 0 \,. \tag{2.135}$$

Since u vanishes at the origin, integration by parts yields

$$\lim_{R\to\infty} \sum_{l'} [u^{J*}_{l'sl''s}(R) u^{J'}_{l'sls}(R) - u^J_{l'sls}(R) u^{J*}_{l'sl''s}(R)] = 0 \,. \tag{2.136}$$

The asymptotic behaviour of Ψ or u, however, is given in (2.110) and is determined by the S-matrix elements. Therefore (2.136) imposes certain conditions on S which are easily shown to be just (2.126).

The properties of symmetry and unitarity reduce the number of parameters necessary to parametrise the S-matrix elements. For total angular momentum $J = 0$ there are two possibilities:

$$l = s = 0 \quad \text{or} \quad l = s = 1 \,.$$

Clearly parity conservation forbids a transition from $l = 0$ to $l = 1$. Therefore the unitarity relation reduces to

$$|S^{J=0}_{ll,\,ll}|^2 = 1, \quad l = 0, 1 \,. \tag{2.137}$$

As a consequence the complex number $S^{J=0}_{ll,\,ll}$ can be parametrised by one real number, the phase $\delta^{J=0}_{ll}$:

$$S^{J=0}_{ll,\,ll} \equiv \exp(2i\,\delta^{J=0}_{ll}) \,. \tag{2.138}$$

This is of course always the case if a partial-wave state is uncoupled. In the usual spectroscopic notation the partial-wave states are denoted by $^{2s+1}l_J$. Therefore the two phases introduced in (2.138) belong to the states $^1S_0$ and $^3P_0$.

For $J > 0$ the S-matrix is fourdimensional. $l = J$ cannot couple to $l = J\pm 1$ because of parity conservation. Therefore one has the combinations $s = 0$ and $s = 1$ for $l = J$, and $s = 1$ for $l = J+1$ and $l = J-1$, and the S-matrix will have the structure

$$S^{J\ne0}_{l'sls} = \begin{pmatrix} S^J_{J-1\,1\,J-1\,1} & S^J_{J-1\,1\,J+1\,1} & 0 & 0 \\ S^J_{J+1\,1\,J-1\,1} & S^J_{J+1\,1\,J+1\,1} & 0 & 0 \\ 0 & 0 & S^J_{J0\,J0} & 0 \\ 0 & 0 & 0 & S^J_{J1\,J1} \end{pmatrix} \,. \tag{2.139}$$

The uncoupled elements are again parametrised by one real phase, whereas the unitary and symmetric 2×2 submatrix needs 3 parameters:

$$S = \begin{pmatrix} \cos 2\bar{\varepsilon} \exp(2i\bar{\delta}_1) & i\sin 2\bar{\varepsilon} \exp[i(\bar{\delta}_1 + \bar{\delta}_2)] \\ i\sin 2\bar{\varepsilon} \exp[i(\bar{\delta}_1 + \bar{\delta}_2)] & \cos 2\bar{\varepsilon} \exp[2i\bar{\delta}_2] \end{pmatrix}. \tag{2.140}$$

This is the "Stapp"- or "bar"-phase shift parametrisation [2.13]. Another parametrisation has been proposed by Blatt and Biedenharn [2.14]. They exploit the fact that S, as an unitary matrix, can be diagonalised by an unitary transformation U:

$$S = U^{-1} \begin{pmatrix} e^{2i\delta_-} & 0 \\ 0 & e^{2i\delta_+} \end{pmatrix} U \tag{2.141}$$

with

$$U = \begin{pmatrix} \cos \varepsilon & \sin \varepsilon \\ -\sin \varepsilon & \cos \varepsilon \end{pmatrix}. \tag{2.142}$$

This yields another 3 parameter form:

$$S = \begin{pmatrix} \cos^2\varepsilon\, e^{2i\delta_-} + \sin^2\varepsilon\, e^{2i\delta_+} & \frac{1}{2}\sin 2\varepsilon(e^{2i\delta_-} - e^{2i\delta_+}) \\ \frac{1}{2}\sin 2\varepsilon(e^{2i\delta_-} - e^{2i\delta_+}) & \cos^2\varepsilon\, e^{2i\delta_+} + \sin^2\varepsilon\, e^{2i\delta_-} \end{pmatrix}. \tag{2.143}$$

It is an easy exercise to work out the relation between the Blatt-Biedenharn and Stapp-parametrisations. One finds

$$\bar{\delta}_1 + \bar{\delta}_2 = \delta_- + \delta_+$$

$$\sin(\bar{\delta}_1 - \bar{\delta}_2) = \tan 2\bar{\varepsilon}/\tan 2\varepsilon \tag{2.144}$$

$$\sin(\delta_- - \delta_+) = \sin 2\bar{\varepsilon}/\sin 2\varepsilon.$$

Exercise: Verify (2.144).

This is as far as one can go without a dynamical calculation. Both experiment and theory have to agree on these phase shift-parameters. The approach from the experimental side is called a phase shift analysis. Since the observables are quadratic forms in the S-matrix elements, a straightforward inversion for the phase parameters is not possible. One works by trial and error to find a set of phases which inserted into the quadratic forms yield the experimentally measured quantities. In this procedure one tries to reduce ambiguities by assuming that the two-nucleon interaction in certain higher partial-wave states is dominated by the one-pion-exchange, which can be calculated with some confidence. We refer the reader for instance to [2.13, 2.15] for more details. Since the "experimental phase shift parameters" are of fundamental importance in characterising the nuclear dynamics we show a few of them [2.16] in the Fig. 2.5a – c. The curves are the energy dependent-analysis phase parameters. We recognize that the important phase shifts in the energy domain below the pion threshold, which is important for nuclear physics, belong to the states $^1S_0(t = 1)$ and $^3S_1 - {}^3D_1(t = 0)$. The parameter ε_1, induced by the tensor force, couples the states $^3S_1$ and $^3D_1$ and plays an important quantitative role for the binding energy of nuclei. Unfortunately, the *np* observables measured up to now are not sensitive enough to pin down the ε_1-value, and one is left with uncomfortably large error bars. The error bars

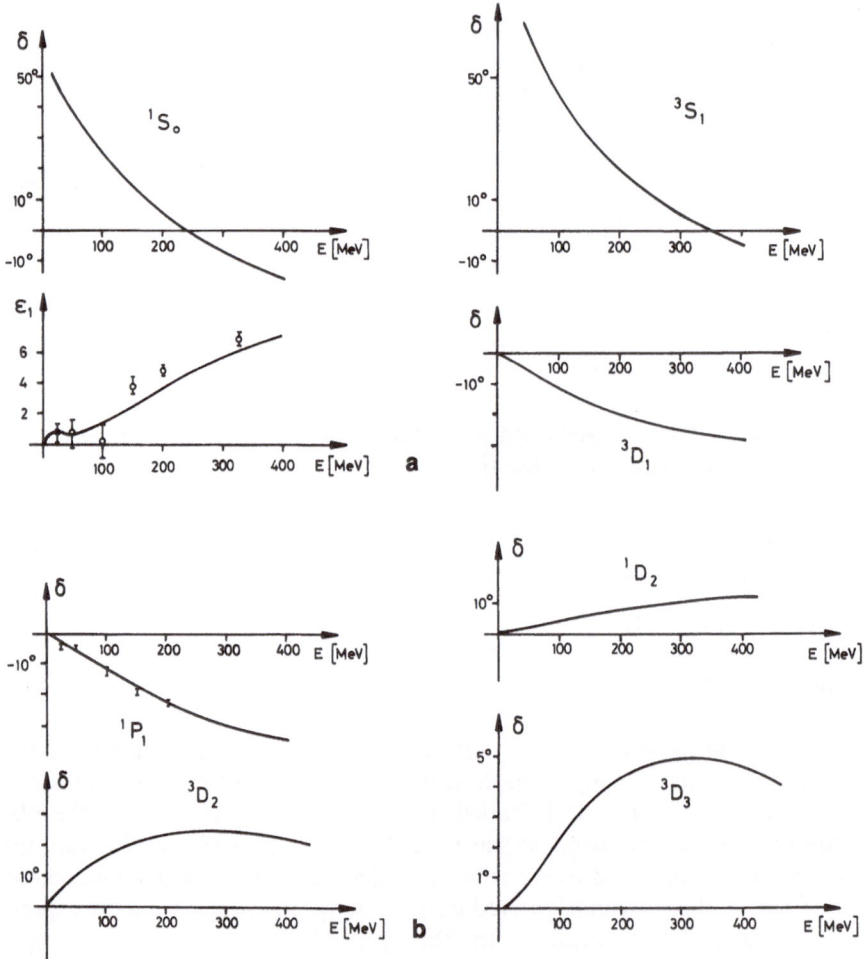

Fig. 2.5a – c. Two nucleon phase shift parameters from an energy dependent analysis [2.16]. Only the cases ε_1 and δ_{1P_1} appear to be not yet settled and error bars are shown. (**a**) Phase parameters for the two-nucleon states, which dominate low energy nuclear physics. (**b**, **c**) *p*- and *d*-wave phase shifts

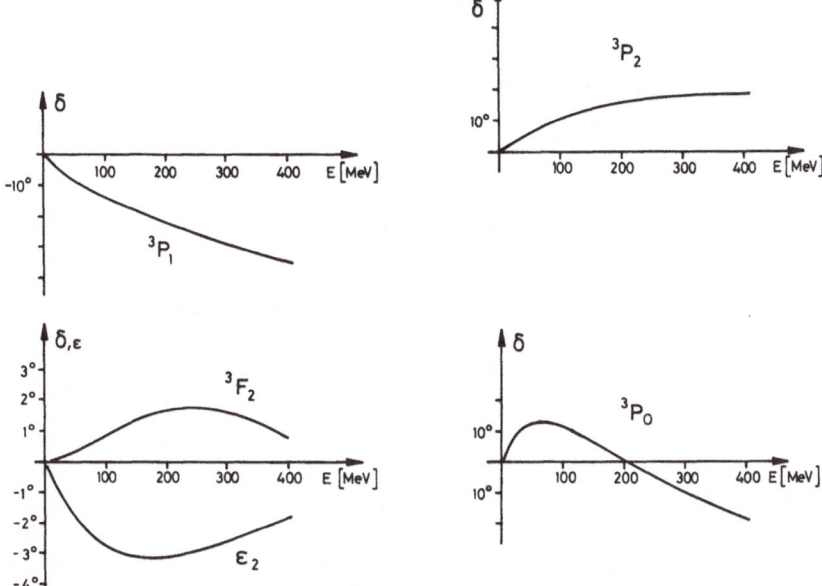

Fig. 2.5c. Figure caption see opposite page

for ε_1 and δ_{1P_1} center around results found in a single energy phase shift analysis. The error bars for the other phase parameters are small on the scale of the figure and are not indicated.

The theoretical calculation of the phases has been a long standing problem and is still essentially unsolved.

2.7 Numerical Methods

Last but not least we want to present a few methods for calculating the phase shift parameters once a potential V is given. We shall regard both coordinate- and momentum-space techniques. The conceptually interesting Padé summation of the Neumann series, even in case it is divergent, will also be described.

2.7.1 Coordinate Space

The coupled set of differential equations is (2.132). As we saw in the previous section, for a given J this reduces either to an uncoupled equation or at most to two coupled equations. Since they are of second order, the number of linearly independent solutions, which are regular at the origin, is equal to the number of equations. Let us regard the case of two coupled equations. Then the physical solution can be presented as a linear combination of the two regular solutions $\Phi_{l's}^{(1)}$ and $\Phi_{l's}^{(2)}$:

$$\Psi_{l's\,ls}^{J}(x) = \alpha_{ls}\,\Phi_{l's}^{(1)}(x) + \beta_{ls}\,\Phi_{l's}^{(2)}(x) \,. \tag{2.145}$$

According to (2.110) $\Psi(x)$ behaves outside the range of $\mathscr{V}$ as

$$\Psi_{l's\,ls}^{J}(x) = \frac{-1}{2i}\,[\delta_{l'l}h_{l'}^{(2)}(qx) - S_{l's\,ls}^{J}h_{l'}^{(1)}(qx)] \,, \tag{2.146}$$

where the spherical Hankel functions have the asymptotic limits:

$$h_{l}^{(1,2)}(qx) \to \frac{\exp[\pm i(qx - \frac{1}{2}l\pi)]}{qx} \,. \tag{2.147}$$

For a given choice of initial quantum numbers ls there are four complex unknowns, α_{ls}, β_{ls}, and $S_{l's\,ls}^{J}$ for the two values of l'. In the case of a single equation there is only one regular solution and there are two complex unknowns, α and S. Since Ψ obeys a second order differential equation it has to be continuous in Ψ and Ψ'. Therefore requiring continuity of the forms (2.145) and (2.146) for Ψ and Ψ' at a radius x beyond the range of $\mathscr{V}$ provides four equations in the case of two coupled equations and two equations in the case of a single equation. In practice equating the two forms (2.145) and (2.146) at two points outside the range of $\mathscr{V}$ provides a convenient set of equations to determine the unknowns. For instance, by eliminating α for the case of a single equation, one easily finds

$$S = \frac{\Phi_{ls}(x_2)\,h_l^{(2)}(qx_1) - \Phi_{ls}(x_1)\,h_l^{(2)}(qx_2)}{\Phi_{ls}(x_2)\,h_l^{(1)}(qx_1) - \Phi_{ls}(x_1)\,h_l^{(1)}(qx_2)} \,, \tag{2.148}$$

which expresses the desired S-matrix element in terms of the numerically determined regular solution Φ and the analytically known spherical Hankel functions $h_l^{(1,2)}$ at two points x_1 and x_2.

Exercise: Derive the expression corresponding to (2.148) for the two-dimensional S-matrix.

In the case of a two-body bound state, the deuteron for instance, the wave function has to decrease exponentially. In other words, the momentum q has to be purely imaginary and therefore the first term (the incoming one) in (2.146) has to be absent. Now both regular solutions $\Phi_{l's}^{(i)}$, $i = 1, 2$ have exponentially increasing and decreasing parts outside the range of $\mathscr{V}$ and for $E = (i|q|)^2 < 0$:

$$\Phi_{l's}^{(i)} \to a_{l's}^{(i)}h_{l'}^{(2)}(i|q|x) + b_{l's}^{(i)}h_{l'}^{(1)}(i|q|x) \,. \tag{2.149}$$

Therefore the linear combination for the physical state behaves outside the range of $\mathscr{V}$ as

$$\Psi^J_{l's}(x) = (\alpha a^{(1)}_{l's} + \beta a^{(2)}_{l's})\, h^{(2)}_{l'}(i|q|x)$$
$$+ (\alpha b^{(1)}_{l's} + \beta b^{(2)}_{l's})\, h^{(1)}_{l'}(i|q|x)\,. \tag{2.150}$$

Consequently, the square integrability condition is

$$\alpha a^{(1)}_{l's} + \beta a^{(2)}_{l's} = 0 \tag{2.151}$$

or denoting the two l'-values by l_1 and l_2 it is

$$\begin{vmatrix} a^{(1)}_{l_1 s} & a^{(2)}_{l_1 s} \\ a^{(1)}_{l_2 s} & a^{(2)}_{l_2 s} \end{vmatrix} = 0\,. \tag{2.152}$$

Of course the coefficients a have to be determined by again matching the form (2.149) to the numerically determined Φ's at two points outside the range of $\mathscr{V}$.

In the case of a large binding energy the numerical procedure requires less care if one integrates the differential equation(s) from inside towards about the middle of the range of $\mathscr{V}$, and also from outside inwards and matches the two numerically determined forms. In this case, two linearily-independent exponentially decaying solutions have to be determined numerically, out of which the physical state is built up analogous to (2.145).

There remains the central question, how does one determine numerically two linearily-independent solutions? A very effective way is the Numerov-method. In order to cover both the uncoupled and coupled cases we introduce a matrix notation

$$\underset{\sim}{\psi}^{(ls)}(x) \equiv (\Psi^J_{l's\,ls}(x)) \equiv \frac{1}{x}\, \underset{\sim}{u}^{(ls)}(x)\,. \tag{2.153}$$

Then the set (2.132) has the form

$$\frac{d^2}{dx^2}\, \underset{\sim}{u}^{(ls)}(x) = \underset{\approx}{\mathscr{F}}(x)\, \underset{\sim}{u}^{(ls)}(x)\,, \tag{2.154}$$

where the matrix $\underset{\approx}{\mathscr{F}}$ contains the centrifugal and energy terms and the potential $\mathscr{V}$. In case of a bound state there is of course no incoming state and therefore no dependence on (ls). By expanding $\underset{\sim}{u}(x)$ in a taylor series and repeatedly using (2.154) to eliminate second derivatives one can establish the following recurrence relation [2.17]

$$\left[1 - \frac{h^2}{12} \underset{\approx}{\mathscr{F}}(1) \right] \underset{\sim}{u}(1) = \left[2 + \frac{5}{6} h^2 \underset{\approx}{\mathscr{F}}(0) \right] \underset{\sim}{u}(0)$$

$$- \left[1 - \frac{h^2}{12} \underset{\approx}{\mathscr{F}}(-1) \right] \underset{\sim}{u}(-1) \tag{2.155}$$

between neighbouring points separated by the distance h. Introducing the auxiliary quantity

$$\underset{\sim}{\xi}(x) \equiv \left[1 - \frac{h^2}{12} \underset{\approx}{\mathscr{F}}(x) \right] \underset{\sim}{u}(x) \tag{2.156}$$

one gets the more convenient form

$$\underset{\sim}{\xi}(i+2) = \left\{ 2 + h^2 \underset{\approx}{\mathscr{F}}(i+1) \left[1 - \frac{h^2}{12} \underset{\approx}{\mathscr{F}}(i+1) \right]^{-1} \right\} \underset{\sim}{\xi}(i+1) - \underset{\sim}{\xi}(i) . \tag{2.157}$$

In case one is interested in the two linearly independent solutions regular at the origin, one can choose the following starting values

$$\underset{\sim}{u}(0) = \underset{\sim}{0}$$

$$\underset{\sim}{u}(h) = \begin{pmatrix} 1 \\ 0 \end{pmatrix} \quad \text{or} \quad \begin{pmatrix} 0 \\ 1 \end{pmatrix} . \tag{2.158}$$

For suitable $\underset{\approx}{\mathscr{F}}$'s these starting values trivially carry over to the ones for $\underset{\sim}{\xi}$. Note however that $l = 1$ is a special case [2.17].

2.7.2 Momentum Space

Momentum space calculations are a valuable alternative to coordinate space treatments and provide independent, reliable tests of the numerics. Moreover, there are cases in which a momentum space treatment is the most natural one, for instance in using meson theoretical potentials which are originally given by Feynman diagrams in momentum space. For treating relativistic equations it even seems compulsory to work in momentum space.

The quantity of interest is $S^J_{l's\,ls}$ which according to (2.111) is determined by the partial-wave projected T-matrix element

$$T^J_{l's\,ls} = \sum_{l''} \langle \mathscr{Y}^{JM}_{l's}j_{l'} | V | \mathscr{Y}^{JM}_{l''s} \Psi^J_{l''s\,ls} \rangle . \tag{2.159}$$

In the same manner as in Sect. 1.5 we can use the Lippmann-Schwinger equations (2.106) for $\Psi^J_{l'sls}$ and the spectral representation of the partial-wave projected free Green's function

$$G_l(xx'E) = \frac{2}{\pi} \int_0^\infty dk\, k^2\, \frac{j_l(kx)j_l(kx')}{E + i\varepsilon - k^2/2\mu} \tag{2.160}$$

to get

$$T^J_{l'sls}(q'q) = V^J_{l'sls}(q'q) + \frac{2}{\pi} \sum_{l''} \int_0^\infty dk\, k^2$$
$$V^J_{l'sl''s}(q'k)\,(E + i\varepsilon - k^2/2\mu)^{-1}\, T^J_{l''sls}(kq)\,. \tag{2.161}$$

The quantities $T^J_{l'sls}(q'q)$

$$T^J_{l'sls}(q'q) = \sum_{l''} \int dx\, x^2 j_{l'}(q'x)\, \mathscr{V}^J_{l'sl''s}(x)\, \Psi^J_{l''sls}(x) \tag{2.162}$$

like the V's in the driving term:

$$V^J_{l'sls}(q'q) = \int_0^\infty dx\, x^2 j_{l'}(q'x)\, \mathscr{V}^J_{l'sls}(x) j_l(qx) \tag{2.163}$$

are half-shell matrix elements since $E = q^2/2\mu \neq q'^2/2\mu$. In the kernel, however, the V's are needed for all momenta q' and k.

As described in Sect. 1.5, it is advantageous to introduce the real K-matrix through

$$K^J_{l'sls}(q'q) = V^J_{l'sls}(q'q) + \frac{2}{\pi} \sum_{l''} \int_0^\infty dk\, k^2\, V^J_{l'sl''s}(q'k)\, P/(E_q - k^2/2\mu) K^J_{l''sls}(kq)\,. \tag{2.164}$$

Then the connection between T and K is simply

$$T^J_{l'sls}(q'q) = \sum_{l''} K^J_{l'sl''s}(q'q)[\delta_{l''l} - 2i\mu q\, T^J_{l''sls}(qq)]\,. \tag{2.165}$$

Putting $q' = q$, we get the on-shell T-matrix required to calculate the S-matrix. On-shell and in matrix notation (2.165) reads

$$T = K(1 - 2i\mu q\, T)\,, \tag{2.166}$$

which can be solved for T:

$$T = (1 + 2i\mu q K)^{-1} K\,. \tag{2.167}$$

Therefore the S-matrix is

$$S = 1 - 4i\mu q T = (1 + 2i\mu q K)^{-1}(1 - 2i\mu q K) , \tag{2.168}$$

which is manifestly unitary, since K is real and symmetric.

To solve (2.164) one must handle the principal-value singularity. One way is to use the trick

$$\int_0^\infty P/(E_q - k^2/2\mu)\,dk = 0 \tag{2.169}$$

and to modify (2.164) to read

$$K_{l's\,ls}^J(q'q) = V_{l's\,ls}^J(q'q) + \frac{2}{\pi} \sum_{l''} \int_0^\infty \frac{dk}{E_q - k^2/2\mu}$$

$$\times [k^2 V_{l's\,l''s}^J(q'k) K_{l''s\,ls}^J(kq) - q^2 V_{l's\,l''s}^J(q'q) K_{l''s\,ls}^J(qq)] . \tag{2.170}$$

Since the bracket vanishes at $k = q$, the singularity is removed and there is no longer any need for a principal-value prescription. Any quadrature rule, like for instance Gauss-Legendre, can be used to discretise the integral. Choosing q' to correspond to the set of quadrature points, we have a system of inhomogeneous algebraic equations. One has as many equations as unknowns if we add another equation for the on-shell value $q' = q$. This method appears to be very convenient and accurate enough for present day nuclear physics.

A more foolproof method numerically is the one proposed by *Kowalski* and *Noyes* [2.18]. They transform the integral equation (2.164) into a Fredholm one. Let us write (2.164) in obvious matrix notation:

$$K(q'q) = V(q'q) + \frac{2}{\pi} \int_0^\infty dk\, k^2 V(q'k) \frac{P}{E_q - k^2/2\mu} K(kq) . \tag{2.171}$$

Define

$$\tau(q'q) = V(q'q) V^{-1}(qq) \tag{2.172}$$

and multiply (2.171) for $q' = q$ by $\tau(q'q)$ from the left. Subtracting the result from (2.171) leads to

$$K(q'q) = \tau(q'q) K(qq) + \frac{2}{\pi} \int_0^\infty dk\, k^2 [V(q'k) - V(q'q) V^{-1}(qq) V(qk)]$$

$$\times \frac{P}{E_q - k^2/2\mu} K(kq) . \tag{2.173}$$

Obviously it is now natural to define the auxiliary matrix $f(q'q)$ through

$$K(q'q) = f(q'q) K(qq) , \tag{2.174}$$

which obeys

$$f(q'q) = \tau(q'q) + \frac{2}{\pi} \int_0^\infty dk \, k^2 [V(q'k)$$

$$- V(q'q) V^{-1}(qq) V(qk)] \, 1/(E_q - k^2/2\mu)^{-1} f(kq) . \tag{2.175}$$

This is the desired integral equation. Since the kernel is nonsingular, one can drop the principal-value prescription. From (2.171) and (2.174) one gets the on-shell K-matrix

$$K(qq) = \left[1 - \frac{2}{\pi} \int_0^\infty dk \, k^2 V(qk) P/(E_q - k^2/2\mu) f(kq) \right]^{-1} V(qq) \tag{2.176}$$

by quadrature. The principal-value integral can be evaluated, for instance using the trick (2.169).

A final remark concerns the direct calculation of the phase shift parameters from the K-matrix elements. Let us solve (2.168) for K:

$$2i\mu q K = (1 - S)(1 + S)^{-1} . \tag{2.177}$$

Then in the Blatt-Biedenharn representation (2.141)

$$S = U^{-1} \beta U$$

with

$$\beta = \begin{pmatrix} e^{2i\delta_-} & 0 \\ 0 & e^{2i\delta_+} \end{pmatrix}$$

we get

$$(1 - \beta)(1 + \beta)^{-1} = 2i\mu q U K U^{-1} . \tag{2.178}$$

From this we find by elementary manipulations

$$\tan \delta_- = -\mu q \left(K_{11} + K_{22} + \frac{K_{11} - K_{22}}{\cos 2\varepsilon} \right)$$

$$\tan \delta_+ = -\mu q \left(K_{11} + K_{22} - \frac{K_{11} - K_{22}}{\cos 2\varepsilon} \right) \tag{2.179}$$

$$\tan 2\varepsilon = \frac{2K_{12}}{K_{11} - K_{22}} .$$

In these formulas K_{ij} denote the elements of the 2×2 K-matrix, $K^J_{l's ls}$.

2.7.3 Padé Method

Let us consider a meromorphic function $f(z)$ which is regular at $z = 0$. The power series expansion

$$f(z) = 1 + a_1 z + a_2 z^2 + \cdots \tag{2.180}$$

converges within a circle which excludes the nearest pole to $z = 0$. However, $f(z)$ can be represented everywhere as

$$f(z) = P(z)/Q(z), \tag{2.181}$$

where P and Q are entire functions:

$$P(z) = 1 + b_1 z + b_2 z^2 + \cdots$$
$$Q(z) = 1 + c_1 z + c_2 z^2 + \cdots. \tag{2.182}$$

The set of coefficients $\{a\}$ on the one side and the sets $\{b\}$ and $\{c\}$ on the other side are related to each other, since they represent the same function. If the latter group could be determined from the first, one could calculate $f(z)$ via (2.181), even outside the circle of convergence. The sequence of Padé approximants serves to achieve that goal. The Padé approximant of order $[N, M]$ is defined by [2.19]

$$f(z) \equiv \frac{P_N(z)}{Q_M(z)} + O(z^{N+M+1}), \tag{2.183}$$

where P_N and Q_M are polynomials of order N and M, respectively, and the Taylor series expansion of $P_N(z)/Q_M(z)$ has to agree with that of $f(z)$ up to the order $Q(z^{N+M})$. The ansatz (2.183) is unique, which is simply shown as follows. Assume there would be a second ratio. Then

$$P'_N/Q'_M = P_N/Q_M + O(z^{N+M+1}) \tag{2.184}$$

or

$$P'_N Q_M = P_N Q'_M + O(z^{N+M+1}). \tag{2.185}$$

It follows that

$$P'_N Q_M = P_N Q'_M \tag{2.186}$$

since the correction term is of higher order.

Four our purposes we only need the theorem that the Padé approximants

$$f_{[N,M]}(z) = \frac{P_N(z)}{Q_M(z)} \tag{2.187}$$

converge towards $f(z)$ for $N, M \to \infty$, if $f(z)$ is a meromorphic function. Unfortunately, this holds only up to a set of points of measure zero [2.20]. In practice, however, this does not seem to cause problems.

How can we apply this theorem to the scattering problems we have met up to now? The Neumann series for the t-operator (1.174), if it converges, is a solution to the Lippmann-Schwinger equation (1.128). It is of the form

$$t(z, \lambda) = V[1 + \lambda K(z) + \lambda^2 K^2(z) + \cdots], \tag{2.188}$$

where $K(z) = G_0(z) V$ and λ is an auxiliary strength parameter, which has to be put equal to 1 for the real physical problem. Obviously the series (2.188) has the formal solution

$$t(z, \lambda) = V[1 - \lambda K(z)]^{-1}. \tag{2.189}$$

We saw in Sect. 1.7 that the kernel $K(z)$ $(\mathrm{Im}\{z\} \neq 0)$ for potential scattering is of the Hilbert-Schmidt type and has a discrete spectrum of eigenvalues $\{\eta_\nu(z)\}$. Therefore $t(z, \lambda)$ given in (2.189) will have poles at $\lambda_\nu = 1/\eta_\nu(z)$, which more-over accumulate only at $\lambda = \infty$. Consequently, $t(z, \lambda)$ is a meromorphic function of λ for each fixed z, and the power series (2.188) will converge for $|\lambda| < \min |1/\eta_\nu(z)|$. Again we see, that whenever an eigenvalue η_ν is outside the unit circle the Neumann series (2.188) will diverge for the physical value $\lambda = 1$. Standard theorems for a Fredholm kernel, like the meromorphic property of its resolvent, are described for instance in [2.21]. The application of the Padé technique is now obvious. Out of the terms of the Neumann series for $\lambda = 1$ one calculates the Padé approximants, which have to converge towards the exact expression (2.189), even if the Neumann series is badly diverging. We shall give an example below.

How does one determine the polynomials in the Padé approximants out of the terms in the Neumann series? Let us regard the diagonal approximants, $N = M$, as an example. They are defined by

$$\frac{1 + b_1 z + \cdots b_N z^N}{1 + c_1 z + \cdots c_N z^N} = 1 + a_1 z + a_2 z^2 + \cdots + a_{2N} z^{2N} + O(z^{2N+1}). \tag{2.190}$$

This is equivalent to

$$(1 + b_1 z + \cdots + b_N z^N) = (1 + c_1 z + \cdots + c_N z^N)(1 + a_1 z + \cdots + a_{2N} z^{2N})$$
$$+ O(z^{2N+1}). \tag{2.191}$$

Equating equal powers in z gives two sets of algebraic equations:

$$
\left.\begin{aligned}
c_N a_1 + c_{N-1} a_2 + \cdots c_1 a_N &= -a_{N+1} \\
c_N a_2 + c_{N-1} a_3 + \cdots c_1 a_{N+1} &= -a_{N+2} \\
\vdots \\
c_N a_N + c_{N-1} a_{N+1} + \cdots c_1 a_{2N-1} &= -a_{2N}
\end{aligned}\right\}
\tag{2.192}
$$

and

$$
\left.\begin{aligned}
c_1 + a_1 &= b_1 \\
c_2 + c_3 a_1 + a_2 &= b_2 \\
\vdots \\
c_N + c_{N-1} a_1 + c_{N-2} a_2 + \cdots a_N &= b_N .
\end{aligned}\right\}
\tag{2.193}
$$

The first set determines the c's, the second the b's.

In practice it may be more convenient to determine special Padé approximants from a continued fraction expansion [2.22]. This is defined by

$$
f(z) = 1 + \cfrac{\alpha_1 z}{1 + \cfrac{\alpha_2 z}{1 + \cfrac{\alpha_3 z}{1 + \cdots}}} .
\tag{2.194}
$$

Putting $\alpha_n = 0$ we get obviously a ratio of polynomials, which has to be a Padé approximant because of its uniqueness. Therefore one faces the question of how to determine the set $\{\alpha_\nu\}$ from the set $\{a_\nu\}$ given in (2.180). First one recognizes that the continued fraction expansion can be generated by the following recursive prescription:

$$
f(z) = f_1(z)
$$
$$
f_n(z) = 1 + \alpha_n z / f_{n+1}(z) .
\tag{2.195}
$$

One can linearise these recurrence relations by putting

$$
f_n(z) = \frac{u_n(z)}{u_{n+1}(z)} .
\tag{2.196}
$$

This yields

$$\frac{u_n(z)}{u_{n+1}(z)} = 1 + \frac{\alpha_n z u_{n+2}(z)}{u_{n+1}(z)} \tag{2.197}$$

or

$$u_n(z) = u_{n+1}(z) + \alpha_n z u_{n+2}(z) . \tag{2.198}$$

Choosing

$$u_n(0) = 1$$
$$u_2(z) \equiv 1 \tag{2.199}$$

the recurrence relations (2.198) can be solved in the following manner:

$n = 1$

$$u_1(z) - u_2(z) = f(z) - 1 = a_1 z + a_2 z^2 + \cdots = \alpha_1 z u_3(z) .$$

It follows from the first requirement in (2.199), that

$$\alpha_1 = a_1$$
$$u_3(z) = 1 + \frac{a_2}{a_1} z + \frac{a_3}{a_1} z^2 + \cdots .$$

$n = 2$

$$u_2(z) - u_3(z) = - \frac{a_2}{a_1} z - \frac{a_3}{a_1} z^2 \cdots = \alpha_2 z u_4 .$$

In the same manner,

$$\alpha_2 = - \frac{a_2}{a_1}$$

$$u_4(z) = 1 + \frac{a_3}{a_2} z + \frac{a_4}{a_2} z^2 + \cdots .$$

$n = 3$

$$u_3(z) - u_4(z) = \left(\frac{a_2}{a_1} - \frac{a_3}{a_2} \right) z + \left(\frac{a_3}{a_1} - \frac{a_4}{a_2} \right) z^2 + \cdots = \alpha_3 z u_5 .$$

Again,

$$\alpha_3 = \frac{a_2}{a_1} - \frac{a_3}{a_2}$$

$$u_5 = 1 + \cfrac{\dfrac{a_3}{a_1} - \dfrac{a_4}{a_2}}{\dfrac{a_2}{a_1} - \dfrac{a_3}{a_2}} z + \cdots$$

etc. Obviously this algorithm lends itself to direct use in a computer program. Once the coefficients α_ν of the continued fraction expansion are known, up to a certain index, the corresponding polynomials in the Padé approximant can be determined recursively. The Padé approximants resulting in this manner are in turn of the orders $[0,0]$, $[1,0]$, $[1,1]$, $[2,1]$, $[2,2]$, which we shall denote just by

$$f_n^{\text{Padé}} \equiv P_n/Q_n, \quad n = 1, 2, \dots. \tag{2.200}$$

Note that the indices in the polynomials $P_n(Q_n)$ have a different meaning as those in (2.183). In lowest order one has

$$P_1 = Q_1 = 1. \tag{2.201}$$

Introducing in addition

$$P_0 = 1$$
$$Q_0 = 0 \tag{2.202}$$

we can write the first nontrivial case with

$$\alpha_1 \neq 0, \quad \alpha_2 = 0 \qquad \text{as} \tag{2.203}$$
$$f_2^{\text{Padé}} = 1 + \alpha_1 z = \frac{P_1 + \alpha_1 z P_0}{Q_1 + \alpha_1 z Q_0}$$

and from (2.200) we have

$$P_2 = P_1 + \alpha_1 z P_0 = 1 + \alpha_1 z$$
$$Q_2 = Q_1 + \alpha_1 z Q_0 = 1. \tag{2.204}$$

This now leads to a recursive construction of higher order Padé approximants. Take $\alpha_2 \neq 0$ but $\alpha_3 = 0$. Then replacing α_1 by $\alpha_1/(1 + \alpha_2 z)$ in (2.203), and using (2.204) we have

$$f_3^{\text{Padé}} = \frac{P_1 + \dfrac{\alpha_1 z}{1 + \alpha_2 z} P_0}{Q_1 + \dfrac{\alpha_1 z}{1 + \alpha_2 z} Q_0} = \frac{P_1 + \alpha_1 z P_0 + \alpha_2 z P_1}{Q_1 + \alpha_1 z Q_0 + \alpha_2 z Q_1}$$

$$= \frac{P_2 + \alpha_2 z P_1}{Q_2 + \alpha_2 z Q_1} \equiv \frac{P_3}{Q_3} . \tag{2.205}$$

We immediately identify

$$P_3 = P_2 + \alpha_2 z P_1$$
$$Q_3 = Q_2 + \alpha_2 z Q_1 . \tag{2.206}$$

By induction the general expression for the Padé approximant, keeping the α's up to $\alpha_n \neq 0$, is

$$P_{n+1} = P_n + \alpha_n z P_{n-1}$$
$$Q_{n+1} = Q_n + \alpha_n z Q_{n-1} , \tag{2.207}$$

where the starting values are (2.201) and (2.202).

As an example of this technique, we apply it to the Neumann series of the t-matrix in the partial wave state $^1S_0$ using the Reid potential (1.172). As we see from Fig. 1.2a there are several eigenvalues outside the unit circle and the series diverges badly. Once the on-shell t-matrix element is given one can calculate the scattering phase shift through (2.111, 159) and (2.138), simplified to

Table 2.1. The nth order terms for the on-shell t-matrix of the Neumann series (2nd column) in comparison with the Padé approximants (3rd column). The resulting phase shift is shown in the 4th column ($E = 12$ MeV)

2	$-0.9374 \times 10^1 - i\,0.4270 \times 10^0$	$-8.966 - i\,0.4270$	0.7983
3	$0.1439 \times 10^3 + i\,0.1960 \times 10^2$	$-0.1608 + i\,0.02215$	1.889
4	$-0.2250 \times 10^4 - i\,0.5262 \times 10^3$	$-0.3871 - i\,0.02685$	2.151
5	$0.3497 \times 10^5 + i\,0.1165 \times 10^5$	$-0.2241 - i\,0.04308$	2.058
6	$-0.5387 \times 10^6 - i\,0.2362 \times 10^6$	$-0.3356 - i\,0.1066$	2.227
7	$0.8218 \times 10^7 + i\,0.4540 \times 10^7$	$-0.1946 - i\,0.2123$	0.8288
8	$-0.1240 \times 10^9 - i\,0.8414 \times 10^8$	$-0.1965 - i\,0.2445$	0.9079
9	$0.1849 \times 10^{10} + i\,0.1518 \times 10^{10}$	$-0.1937 - i\,0.2239$	0.8586
10	$-0.2717 \times 10^{11} - i\,0.2683 \times 10^{11}$	$-0.1932 - i\,0.2248$	0.8610
11	$0.3928 \times 10^{12} + i\,0.4660 \times 10^{12}$	$-0.1932 - i\,0.2246$	0.8604
12	$-0.5564 \times 10^{13} - i\,0.7980 \times 10^{13}$	$-0.1932 - i\,0.2247$	0.8607
13	$0.7680 \times 10^{14} + i\,0.1350 \times 10^{15}$	$-0.1932 - i\,0.2246$	0.8606
14	$-0.1204 \times 10^{16} - i\,0.2257 \times 10^{16}$	$-0.1932 - i\,0.2247$	0.8607

the uncoupled case. In Table 2.1 we show the first few terms of the Neumann series for the on-shell t-matrix element together with the corresponding Padé approximants. The through column contains the phase shifts resulting from the Padé approximants. The direct inversion of the Lippmann-Schwinger equation, using the same quadrature points, yields $\delta = 0.8607$ in perfect agreement.

3. Three Interacting Particles

One is taught to think in terms of single particle motions. We studied some cases in the last two chapters, where this is strictly correct. A three-particle system involves two relative motions, which are not independent from each other, and simpler dynamical pictures can be only approximate. We shall introduce various rigorous formulations of the three-body system, which will clarify how the increase in complexity of the motions necessarily requires an extended mathematical apparatus. The essential new dynamical feature is the occurrence of rearrangement channels and of the break-up channel. This requires more boundary conditions to determine a scattering state than are necessary in a two-body system. Through the techniques presented are fairly general, the applications described refer solely to nucleons.

3.1 Channels

Three particles may interact in general by two- and three-body forces. Let us assume that all interactions are of finite range. Then beyond a certain distance away from the center of mass, the forces *between all three particles* will have dropped to zero, although two-body forces between a pair of particles may still be active depending on the geometrical configuration. The three particles can obviously form various arrangements. Two particles can be in a bound state with the third particle well separated from the pair. This is called a two-body fragmentation channel. Since the two fragments do not interact, the type of fragmentation cannot change any more as the fragments separate further. In general there will be three types of two-body fragmentation channels, which we denote in an obvious notation by

$$1,23; \quad 2,31; \quad 3,12. \tag{3.1a}$$

The remaining arrangement is that no pair is bound, which is called the three-body break-up channel:

$$1,2,3. \tag{3.1b}$$

In the asymptotic limit in that channel, all three particles are well separated from each other and they move freely. There is one very interesting exception, however, where two particles leave with equal or nearly equal momenta and will therefore interact with each other for a very long time. One speaks in this case of a final-state interaction in the break-up channel.

Through each of these 4 channels a scattering process can be initiated, which may lead in general to all four possible exit channels again. For instance one may have the following transitions:

initial channel   final channels

$$1 \rightarrow \begin{pmatrix} 2 \\ 3 \end{pmatrix} \rightarrow \begin{cases} 1,23 & \text{elastic} \\ 2,31 & \text{rearrangement} \\ 3,12 & \text{rearrangement} \\ 1,23 & \text{break-up ,} \end{cases} \tag{3.2}$$

which comprises elastic, rearrangement and break-up processes. If the pair interaction supports various bound states the bound pairs can live in any of them.

It is customary to designate the two-body fragmentation channel by the the single particle. Thus channel 1 denotes the arrangements $1,23$, etc. It is also natural to introduce three sets of Jacobi coordinates

$$\begin{vmatrix} r_1 = x_2 - x_3 & r_2 = x_3 - x_1 & r_3 = x_1 - x_2 \\ R_1 = x_1 - \frac{1}{2}(x_2 + x_3) & R_2 = x_2 - \frac{1}{2}(x_3 + x_1) & R_3 = x_3 - \frac{1}{2}(x_1 + x_2) \end{vmatrix}, \tag{3.3a}$$

which are convenient for describing the relative motions in channels 1, 2, and 3, respectively. (These sets refer to equal mass particles.) Obviously one pair (r_k, R_k) describes completely the relative motions, and the other two pairs (r_l, R_l), $l \neq k$, can be expressed linearly in terms of (r_k, R_k).

Let k_i, $i = 1, 2, 3$ be the individual momenta of the three particles and $K = \sum k_i$ the total momentum. Then the relative momenta (p_l, q_l), conjugate to (r_l, R_l), give for the kinetic energy

$$H_0 = \sum \frac{k_i^2}{2m} = \frac{K^2}{2M} + \frac{p_l^2}{2\mu_l} + \frac{q_l^2}{2M_l}, \quad l = 1, 2, 3 \tag{3.4}$$

and are related to the k_i by

$$\begin{vmatrix} p_1 = \frac{1}{2}(k_2 - k_3) & p_2 = \frac{1}{2}(k_3 - k_1) & p_3 = \frac{1}{2}(k_1 - k_2) \\ q_1 = \frac{2}{3}[k_1 - \frac{1}{2}(k_2 + k_3)] & q_2 = \frac{2}{3}[k_2 - \frac{1}{2}(k_3 + k_1)] & q_3 = \frac{2}{3}[k_3 - \frac{1}{2}(k_1 + k_2)] \end{vmatrix}.$$

$$\tag{3.3b}$$

For equal mass particles the various masses are $M = 3m$, $\mu_l = \frac{1}{2}m$, $M_l = \frac{2}{3}m$. In the following we put $m = 1$. Again one pair describes completely the relative motion in momentum space. We find for instance that

$$p_2 = -\tfrac{1}{2}p_1 - \tfrac{3}{4}q_1 \qquad p_3 = -\tfrac{1}{2}p_1 + \tfrac{3}{4}q_1$$
$$q_2 = p_1 - \tfrac{1}{2}q_1 \qquad\quad q_3 = -p_1 - \tfrac{1}{2}q_1 . \tag{3.5}$$

Remaining relations follow by cyclic permution of (3.5).

Aside from internal degrees of freedom the momentum states $|k_1 k_2 k_3\rangle$ span the Hilbert space for the three particles:

$$\int dk_1 dk_2 dk_3 |k_1 k_2 k_3\rangle \langle k_1 k_2 k_3| = 1 . \tag{3.6a}$$

Then in accordance with (3.3b) we introduce states of relative motion by

$$\langle k_1 k_2 k_3 | p_k q_k K\rangle \equiv \delta(p_k - \tfrac{1}{2}(k_l - k_m)) \, \delta(q_k - \tfrac{2}{3}[k_k - \tfrac{1}{2}(k_l + k_m)])$$
$$\delta(K - k_1 - k_2 - k_3) , \tag{3.6b}$$

where $(k\,l_m)$ is a cyclic permutation of (1 2 3), It is easily verified that these states are again orthonormalised and complete:

$$\langle p'_\alpha q'_\alpha K' | p_\alpha q_\alpha K\rangle = \delta(p'_\alpha - p_\alpha) \, \delta(q'_\alpha - q_\alpha) \, \delta(K' - K) \tag{3.7a}$$

and

$$\int dp_\alpha dq_\alpha dK | p_\alpha q_\alpha K\rangle \langle p_\alpha q_\alpha K| = 1 . \tag{3.7b}$$

These momentum states just describe the free motions and are therefore eigenstates to the operator of kinetic energy, H_0:

$$H_0 | p_\alpha q_\alpha\rangle = (p_\alpha^2 + \tfrac{3}{4}q_\alpha^2) | p_\alpha q_\alpha\rangle \equiv H_0 | \phi_{pq}\rangle \equiv H_0 | \phi_0\rangle . \tag{3.8}$$

Since the total momentum K is conserved, we put it to zero and shall not show it explicitly in the following.

Pair interactions will be denoted by the same simple rule as for the channels, thus $V_1 = V_{23}$ etc. In the channel α, the pair interaction V_α binds two particles while particle α is far away and does not interact. This is described by the so called channel states

$$|\phi_{q_\alpha}\rangle = |\varphi_\alpha\rangle |q_\alpha\rangle . \tag{3.9}$$

Here $|\varphi_\alpha\rangle$ is the bound state pair wave function and $|q_\alpha\rangle$ the state of free motion of particle α with respect to the bound pair. Were the q_α-dependence in the state $|\phi_{q_\alpha}\rangle$ is unimportant, we shall use the notation $|\phi_\alpha\rangle$. This state (3.9) is an eigenstate to the channel Hamiltonian

$$H_\alpha \equiv H_0 + V_\alpha. \tag{3.10}$$

Namely we have

$$H_\alpha|\phi_\alpha\rangle = (\varepsilon_\alpha + \tfrac{3}{4}q_\alpha^2)|\phi_\alpha\rangle \equiv E_{q_\alpha}|\phi_\alpha\rangle, \tag{3.11}$$

where $\varepsilon_\alpha < 0$ is the bound state energy of the state $|\varphi_\alpha\rangle$.

There are other eigenstates of H_α, in which the pair is not bound but in a scattering state. This is of course a break-up configuration with just one pair interaction occurring. These states are obviously

$$|\phi_\alpha\rangle^{(+)} = |p_\alpha\rangle^{(+)}|q_\alpha\rangle \tag{3.12}$$

and obey

$$H_\alpha|\phi_\alpha\rangle^{(+)} = (p_\alpha^2 + \tfrac{3}{4}q_\alpha^2)|\phi_\alpha\rangle^{(+)} \equiv E_{p_\alpha q_\alpha}|\phi_\alpha\rangle^{(+)}. \tag{3.13}$$

The total Hamilton operator results if we add to H_α the interaction of particle α with the pair:

$$V^\alpha \equiv V_\beta + V_\gamma + V_4, \quad \beta \neq \alpha \neq \gamma. \tag{3.14}$$

Besides the two pair interactions, a three-body force V_4 is included which might be important in certain systems. Thus

$$H = H_0 + V_\alpha + V^\alpha = H_\alpha + V^\alpha. \tag{3.15}$$

It is convenient to denote the break-up channel by the index 0 and to introduce

$$V_0 \equiv 0$$
$$V^0 \equiv V + V_4 \equiv \Sigma V_\alpha + V_4. \tag{3.16}$$

Then the second equality in (3.15) can be used for all four channels $\alpha = 0, 1, 2, 3$.

3.2 The Fundamental Set of Lippmann-Schwinger Equations

The scattering process can be initiated through each of the 4 channels described in the last section. In the two body fragmentation channels, the relative motion of particle α towards the bound pair can be localised by a wave packet, exactly as we proceeded in potential scattering. Therefore the time dependent channel state will be

$$|\phi_\alpha(t)\rangle = \int dq_\alpha |\phi_{q_\alpha}\rangle \exp(-iE_{q_\alpha}t) f_0(q_\alpha) , \qquad (3.17a)$$

where the momentum distribution $f_0(q)$ is peaked around q_α^0. Similarily a state of three free particles moving towards each other will be described by wave packets in both relative motions

$$|\phi_0(t)\rangle = \int d^3p\, d^3q\, |\phi_{pq}\rangle \exp(-iE_{pq}t) f_0(p) f_0(q) . \qquad (3.17b)$$

Here p and q stands for any of the pairs (p_α, q_α). Of course one can also incorporate immediately one pair interaction V_α and distort the initial wave packet for three particles as

$$|\phi_\alpha^{(+)}(t)\rangle = \int dp_\alpha dq_\alpha |\phi_\alpha^{(+)}\rangle \exp(-iE_{p_\alpha q_\alpha}t) f_0(p_\alpha) f_0(q_\alpha) . \qquad (3.18)$$

Clearly the states (3.17a) and (3.18) obey

$$H_\alpha \left\{ \begin{array}{c} |\phi_\alpha(t)\rangle \\ |\phi_\alpha^{(+)}(t)\rangle \end{array} \right\} = i\frac{\partial}{\partial t} \left\{ \begin{array}{c} |\phi_\alpha(t)\rangle \\ |\phi_\alpha^{(+)}(t)\rangle \end{array} \right\} \alpha = 1, 2, 3 , \qquad (3.19)$$

whereas the time dependence of $|\phi_0(t)\rangle$ is governed only by the operator of kinetic energy, H_0. Out of the states in the four different channels develop four different solutions $|\Psi_\alpha^{(+)}(t)\rangle$, $\alpha = 0, 1, 2, 3$ of the full Schrödinger equation

$$(H_\alpha + V^\alpha)|\Psi_\alpha^{(+)}(t)\rangle = i\frac{\partial|\Psi_\alpha^{(+)}(t)\rangle}{\partial t} . \qquad (3.20)$$

As in potential scattering they are linked to the initial channel states by the requirements

$$\lim_{t\to-\infty} \| \Psi_\alpha^{(+)}(t) - \phi_\alpha(t) \| = 0 \qquad (3.21)$$

or

$$|\Psi_\alpha^{(+)}(0)\rangle = \lim_{t\to-\infty} e^{iHt} e^{-iH_\alpha t}|\phi_\alpha(0)\rangle . \qquad (3.22)$$

The two equations (3.21, 22) also apply to the case that the initial-state-pair interaction V_α is already included in the break-up channel: $|\phi_\alpha(t)\rangle \to |\phi_\alpha^{(+)}(t)\rangle$.

The proof for the existence of these limits is rather close to the one sketched in Chap. 1. We refer the reader to original papers and a lecture by *Hunziker* [3.1]. We are thus led to introduce four Möller channel operators

$$\Omega_\alpha^{(+)} = \lim_{t \to -\infty} e^{iHt} e^{-iH_\alpha t}, \quad \alpha = 0, 1, 2, 3 , \tag{3.23}$$

which map the 4 types of channel states into 4 types of solutions of the full Schrödinger equation. Note that $|\Psi_0^{(+)}(t)\rangle$ can be reached by applying $\Omega_0^{(+)}$ to $|\phi_0(t)\rangle$ or $\Omega_\alpha^{(+)}$ to $|\phi_\alpha^{(+)}(t)\rangle$. This will be reflected later in the existence of two types of Lippmann-Schwinger equations for the corresponding time independent scattering state.

As in Sect. 1.1, we rewrite the time limit as an ε-limit and find 4 types of stationary scattering states to sharp initial momenta:

$$|\Psi_{q_\alpha}^{(+)}\rangle = \lim_{\varepsilon \to 0} \frac{i\varepsilon}{E_{q_\alpha} + i\varepsilon - H} |\phi_{q_\alpha}\rangle, \quad \alpha = 1, 2, 3 \tag{3.24}$$

$$|\Psi_{pq}^{(+)}\rangle = \lim_{\varepsilon \to 0} \frac{i\varepsilon}{E_{pq} + i\varepsilon - H} |\phi_{pq}\rangle, \quad \alpha = 0 . \tag{3.25}$$

Clearly these 4 types of states will be solutions to the stationary Schrödinger equation. Also they are mutually orthogonal. This can be shown as in Sect. 1.3. The time dependent scattering states have the representations

$$|\Psi_\alpha^{(+)}(t)\rangle = \int d\alpha |\Psi_\alpha^{(+)}\rangle e^{-iE_\alpha t} f_0(\alpha) , \quad \alpha = 0, 1, 2, 3 , \tag{3.26}$$

where the symbol α stands for q_α or pq, respectively. Since the scalar product between the states (3.26) is time independent and because of (3.21) we get

$$\langle \Psi_\alpha^{(+)}(0) | \Psi_\beta^{(+)}(0) \rangle = \lim_{t \to -\infty} \langle \phi_\alpha(t) | \phi_\beta(t) \rangle . \tag{3.27}$$

For $\alpha \neq \beta$ the right hand side vanishes by the very definition of channels, which asymptotically have no overlap. Thus

$$\langle \Psi_\alpha^{(+)}(0) | \Psi_\beta^{(+)}(0) \rangle = \delta_{\beta\alpha} \langle \phi_\alpha(0) | \phi_\beta(0) \rangle \tag{3.28a}$$

or

$$\int d\alpha \, d\beta' \langle \Psi_\alpha^{(+)} | \Psi_{\beta'}^{(+)} \rangle f_0(\alpha) f_0(\beta')$$
$$= \delta_{\beta\alpha} \int d\alpha \, d\alpha' f_0(\alpha) f_0(\alpha') \delta(\alpha - \alpha') . \tag{3.28b}$$

This can only be satisfied if

$$\langle \Psi_\alpha^{(+)} | \Psi_\beta^{(+)} \rangle = 0 \quad \text{for} \quad \alpha \neq \beta \tag{3.29}$$

and

$$\langle \Psi_{q_\alpha}^{(+)} | \Psi_{q'_\alpha}^{(+)} \rangle = \delta^3(q_\alpha - q'_\alpha)$$
$$\langle \Psi_{pq}^{(+)} | \Psi_{pq'}^{(+)} \rangle = \delta^3(p - p') \, \delta^3(q - q') \, . \tag{3.30}$$

Having exhausted all possible initial channels, we assume that the set of states (3.24, 25), together with three-particle bound states, from a complete set in the three-particle Hilbert space. A nice presentation of the mapping properties of the channel Möller operators and the remapping of the adjoint ones in the Hilbert space can be found in [3.2].

Proceeding as in Sect. 1.2 and regarding the amplitude $\langle \phi_\beta(t) | \Psi_\alpha^{(+)}(t) \rangle$, we are led to transition amplitudes from channel α to channel β

$$T_{\beta\alpha} \equiv \langle \phi_\beta | V^\beta | \Psi_\alpha^{(+)} \rangle \, . \tag{3.31}$$

Besides a phase-space factor its absolute square determines the cross section from channel α to channel β. Note the typical structure of $T_{\beta\alpha}$. It contains the final channel state ϕ_β, the interaction V^β between the fragments in channel β and the scattering state of the full stationary Schrödinger equation belonging to the initial channel.

We are now faced with the problem of setting up integral equations for $| \Psi_\alpha^{(+)} \rangle$ and $T_{\beta\alpha}$. A natural idea is to apply resolvent identities as in Chap. 1.2 and to establish Lippmann-Schwinger equations. We introduce the channel resolvent operators

$$G_\alpha(z) \equiv \frac{1}{z - H_\alpha} \tag{3.32}$$

and the full resolvent operator

$$G(z) \equiv \frac{1}{z - H} \, , \tag{3.33}$$

which fulfill the obvious identities

$$G(z) = G_\alpha(z) + G_\alpha(z) \, V^\alpha G(z)$$
$$= G_\alpha(z) + G(z) \, V^\alpha G_\alpha(z) \, . \tag{3.34}$$

Inserting the first equality into (3.24) and (3.25) gives

$$| \Psi_\alpha^{(+)} \rangle = \phi_\alpha + \lim_{\varepsilon \to 0} \frac{1}{E_\alpha + i\varepsilon - H_\alpha} V^\alpha | \Psi_\alpha^{(+)} \rangle \, , \quad \alpha = 0, 1, 2, 3 \tag{3.35}$$

and thus four types of Lippmann-Schwinger equations for the four types of states. Unfortunately this result is not as nice as it may look. In rewriting $\Omega_\alpha^{(+)}$ we have used the resolvent identity with G_α. What will happen if we instead use the one with $G_\beta, \beta \neq \alpha$? In that case we get

$$|\Psi_\alpha^{(+)}\rangle = \lim_{\varepsilon \to 0} \frac{i\varepsilon}{E_\alpha + i\varepsilon - H_\beta} |\phi_\alpha\rangle + \lim_{\varepsilon \to 0} \frac{1}{E_\alpha + i\varepsilon - H_\beta} V^\beta |\Psi_\alpha^{(+)}\rangle, \quad \beta \neq \alpha. \tag{3.36}$$

Let us first regard the case $\alpha \neq 0$, that is two fragments in the initial state. We claim that

$$\lim_{\varepsilon \to 0} \frac{i\varepsilon}{E_\alpha + i\varepsilon - H_\beta} |\phi_\alpha\rangle = 0 \quad (\alpha \neq 0, \beta \neq \alpha). \tag{3.37}$$

This is known as Lippmann's identity [3.3]. Let us illuminate the mechanism on the left hand side. To that aim we need the spectral representation of the resolvent operator G_β. Clearly the eigenstates of H_β given in (3.9) and (3.12) are complete in the space of relative motions and therefore in the three-particle Hilbert space:

$$\int dq_\beta |\phi_{q_\beta}\rangle\langle\phi_{q_\beta}| + \int dp_\beta dq_\beta |\phi_{p_\beta q_\beta}^{(+)}\rangle\langle\phi_{p_\beta q_\beta}^{(+)}| = 1. \tag{3.38}$$

Therefore G_β applied onto ϕ_α can be explicitly written as

$$G_\beta(E_\alpha + i\varepsilon)|\phi_\alpha\rangle = \int dq_\beta' |\phi_{q_\beta'}\rangle \frac{1}{E_\alpha + i\varepsilon - \frac{3}{4}q_\beta'^2} \langle\phi_{q_\beta'}|\phi_\alpha\rangle$$

$$+ \iint dp_\beta' dq_\beta' |\phi_{p_\beta' q_\beta'}^{(+)}\rangle \frac{1}{E_\alpha + i\varepsilon - p_\beta'^2 - \frac{3}{4}q_\beta'^2} {}^{(+)}\langle\phi_{p_\beta' q_\beta'}|\phi_\alpha\rangle. \tag{3.39}$$

The integrals exist in the limit $\varepsilon \to 0$ if the overlap matrix elements are not singular in the integral variables at the values at which the denominator vanishes. For $\beta \neq \alpha$, $\beta \neq 0$, and $\alpha \neq 0$

$$\langle\phi_{q_\beta'}|\phi_\alpha\rangle = \langle\varphi_\beta|\langle q_\beta'|\varphi_\alpha\rangle|q_\alpha\rangle \tag{3.40}$$

is nonsingular in q_β', since the product of two bound state pair wave functions for different pairs have a nonzero overlap only in a finite region of space, and the Fourier transform implied by $\langle q_\beta'|$ yields a smooth nonsingular function in q_β'. Thus the first term in (3.39), multiplied by $i\varepsilon$, vanishes in the limit $\varepsilon \to 0$. The second overlap matrix element is

$$^{(+)}\langle\phi_{p_\beta' q_\beta'}|\phi_\alpha\rangle = {}^{(+)}\langle p_\beta'|\langle q_\beta'|\varphi_\alpha\rangle|q_\alpha\rangle. \tag{3.41}$$

Its most singular part results from the momentum eigenstates $\langle p'_\beta|$, which is part of the two-body scattering state $^{(+)}\langle p'_\beta|$:

$$\langle p'_\beta|\langle q'_\beta|\varphi_\alpha\rangle|q_\alpha\rangle = \langle p'_\alpha|\langle q'_\alpha|\varphi_\alpha\rangle|q_\alpha\rangle$$
$$= \delta^3(q'_\alpha - q_\alpha)\langle p'_\alpha|\varphi_\alpha\rangle. \tag{3.42}$$

In the first equality we used the fact that free states can be represented by different sets of Jacobi momenta. The connection between (p'_β, q'_β) and (p'_α, q'_α) is given in (3.5). This connection also guarantees that

$$p'^2_\beta + \tfrac{3}{4}q'^2_\beta = p'^2_\alpha + \tfrac{3}{4}q'^2_\alpha \tag{3.43}$$

and that the functional determinant for transformation from $(p'_\beta q'_\beta)$ to $(p'_\alpha q'_\alpha)$ is unity. Thus the second integral in (3.39) for the most singular part of (3.41) is

$$\int dp'_\alpha dq'_\alpha |\phi^{(+)}_{p'_\beta q'_\beta}\rangle \frac{1}{E_\alpha + i\varepsilon - p'^2_\alpha - \tfrac{3}{4}q'^2_\alpha} \delta(q_\alpha - q'_\alpha)\langle p'_\alpha|\varphi_\alpha\rangle$$

$$= \int dp'_\alpha |\phi^{(+)}_{p'_\beta q'_\beta}\rangle \frac{1}{E_\alpha + i\varepsilon - \tfrac{3}{4}q^2_\alpha - p'^2_\alpha} \langle p'_\alpha|\varphi_\alpha\rangle. \tag{3.44}$$

Now $E_\alpha - \tfrac{3}{4}q^2_\alpha = \varepsilon_\alpha < 0$ and the denominator cannot vanish. Thus the limit $\varepsilon \to 0$ exists and multiplication by $i\varepsilon$ also makes that part equal to zero in the limit $\varepsilon \to 0$. The discussion carried through in (3.42) and (3.44) obviously covers also the case $\beta = 0$.

We arrive at the puzzling result that the scattering state $|\Psi^{(+)}_\alpha\rangle$, $\alpha \neq 0$, obeys in addition to the Lippmann-Schwinger equation (3.35) the homogeneous equations

$$|\Psi^{(+)}_\alpha\rangle = \lim_{\varepsilon \to 0} G_\beta(E_\alpha + i\varepsilon) V^\beta |\Psi^{(+)}_\alpha\rangle, \quad \beta \neq \alpha. \tag{3.45}$$

One concludes immediately that the Lippmann-Schwinger equation (3.35) does not define $|\Psi^{(+)}_\alpha\rangle$ uniquely, since the corresponding homogeneous equation has nontrivial solutions, namely $|\Psi^{(+)}_\beta\rangle$, $0 \neq \beta \neq \alpha$. This was one of the main reasons for disregarding the Lippmann-Schwinger equations in n-body scattering theory for $n > 2$ [3.4, 5].

It is important to recognize that this problem of nonuniqueness is present only if one works on the real axis, that is if the limit $\varepsilon \to 0$ in the resolvent operators has been taken. Keeping $\varepsilon \neq 0$, Eq. (3.35) defines the state $|\Psi_\alpha(\varepsilon)\rangle$ uniquely:

$$|\Psi_\alpha(\varepsilon)\rangle = |\phi_\alpha\rangle + \frac{1}{E_\alpha + i\varepsilon - H_\alpha} V^\alpha |\Psi_\alpha(\varepsilon)\rangle, \quad \alpha = 0, 1, 2, 3. \tag{3.46}$$

Indeed for $\text{Im}\{z\} \neq 0$ it follows from (3.34) that

$$[1 + G(z) V^\alpha][1 - G_\alpha(z) V^\alpha] = 1 , \tag{3.47}$$

which yields the formal solution

$$|\Psi_\alpha(\varepsilon)\rangle = |\phi_\alpha\rangle + G(E_\alpha + i\varepsilon) V^\alpha |\phi_\alpha\rangle \tag{3.48}$$

expressed in terms of the full resolvent operator. Since the transition from (3.24) to (3.36) is just an operator identity for $\varepsilon \neq 0$

$$|\Psi_\alpha(\varepsilon)\rangle = \frac{i\varepsilon}{E_\alpha + i\varepsilon - H_\beta} |\phi_\alpha\rangle + \frac{1}{E_\alpha + i\varepsilon - H_\beta} V^\beta |\Psi_\alpha(\varepsilon)\rangle \tag{3.49}$$

also defines $|\Psi_\alpha(\varepsilon)\rangle$ uniquely. Therefore, if a numerically stable algorithm could be found, one could construct the physical solution $|\Psi_\alpha^{(+)}\rangle$ from the sequence of states $|\Psi_\alpha(\varepsilon_n)\rangle$ with ε_n going to zero.

Let us go back to the real axis ($\varepsilon = 0$), and let us insist on understanding the various relations between the inhomogeneous and homogeneous Lippmann-Schwinger equations. This will lead us to a quite transparent basis for the formulation of scattering theory for n-particles. First we notice that the kernel $G_\alpha V^\alpha$ leads necessarily to a purely outgoing wave in channel α. For $\alpha \neq 0$ this is obvious from a glance at the first part on the rhs of (3.39), which shows the spectral decomposition of G_α responsible for the behaviour in channel α. Remembering (3.9) we recognize just the free Green's function (1.60), now for the relative motion of particle α with respect to the bound pair. However, the problem is that the kernel $G_\alpha V^\alpha$ allows not only outgoing waves but also incoming waves for channels $\beta \neq \alpha$. This follows from the fact that $|\Psi_\beta^{(+)}\rangle$, which contains certainly the ingoing waves in $|\phi_\beta\rangle$, is an eigenstate of $G_\alpha V^\alpha$ according to (3.45). The mechanism for the passage of $|\phi_\beta\rangle$ through $G_\alpha V^\alpha$ is clearly that the continuum part of the spectral representation of G_α contains G_0 and it is the part $G_0 V_\beta$ in $G_\alpha V^\alpha$ which is the slit. Namely one has

$$G_0 V_\beta |\phi_\beta\rangle = |\phi_\beta\rangle . \tag{3.50}$$

Exercise: Verify (3.50).

Therefore the Lippmann-Schwinger equation (3.35) does not specify the boundary conditions and cannot single out one specific scattering state. That equation has to be augmented. What do we expect as boundary conditions for $|\Psi_\alpha^{(+)}\rangle$? Certainly the scattered part $(|\Psi_\alpha^{(+)}\rangle - |\phi_\alpha\rangle)$ should be purely outgoing in all 4 channels (if they exist and if they are open). Now $|\Psi_\alpha^{(+)}\rangle$ also obeys the homogeneous equations (3.45) for $\beta \neq \alpha$, which tells us directly (as can be seen from the spectral decomposition of G_β as explained above) that indeed $|\Psi_\alpha^{(+)}\rangle$ is purely outgoing in the remaining two-body fragmentation channels as well. So it appears very natural to impose the boundary conditions in all two-body fragmentation channels by augmenting the Lippmann-Schwinger equation (3.35) with the two homogeneous ones (3.45) ($\alpha\beta\gamma$ out of 1 2 3 and all different):

$$|\Psi_\alpha^{(+)}\rangle = |\phi_\alpha\rangle + G_\alpha V^\alpha |\Psi_\alpha^{(+)}\rangle \qquad (3.51\,\text{a})$$

$$|\Psi_\alpha^{(+)}\rangle = \qquad G_\beta V^\beta |\Psi_\alpha^{(+)}\rangle \qquad (3.51\,\text{b})$$

$$|\Psi_\alpha^{(+)}\rangle = \qquad G_\gamma V^\gamma |\Psi_\alpha^{(+)}\rangle \,. \qquad (3.51\,\text{c})$$

Now we claim that this set (3.51a − c) defines $|\Psi_\alpha^{(+)}\rangle$ uniquely [3.6]. It is indeed a necessary and sufficient set to define $|\Psi_\alpha^{(+)}\rangle$ uniquely. This can be seen in the following manner. Obviously every solution of this set is a solution of the Schrödinger equation. Also we assume on physical grounds that the complete set of solutions is

$$|\Psi_1^{(+)}\rangle, \quad |\Psi_2^{(+)}\rangle, \quad |\Psi_3^{(+)}\rangle, \quad |\Psi_0^{(+)}\rangle, \quad |\Psi_{\text{bound}}\rangle \,. \qquad (3.52)$$

Now take the example $\alpha = 1$, $\beta = 2$, $\gamma = 3$. Then (3.51a) alone would allow arbitrary admixtures of $|\Psi_2^{(+)}\rangle$ and $|\Psi_3^{(+)}\rangle$ on top of the desired solution $|\Psi_1^{(+)}\rangle$. Since $|\Psi_2^{(+)}\rangle$ obeys the inhomogeneous version of (3.51b) with the driving term $|\phi_2\rangle$, $|\Psi_2^{(+)}\rangle$ is excluded if we require (3.51b) in addition to (3.51a). In the same manner the third equation (3.51c) excludes $|\Psi_3^{(+)}\rangle$. Possible three-body bound states, $|\Psi_{\text{bound}}\rangle$, will obey the homogeneous equations, however for different energies, and are therefore trivially excluded. It remains to consider $|\Psi_0^{(+)}\rangle$ which is defined by

$$|\Psi_0^{(+)}\rangle = i\varepsilon\, G(E_{pq} + i\varepsilon)|\phi_{pq}\rangle \,. \qquad (3.53)$$

Inserting the resolvent identity (3.34) we are led for the first term to

$$\lim_{\varepsilon\to 0} \frac{i\varepsilon}{E_{pq} + i\varepsilon - H_0 - V_\alpha}|p_\alpha q_\alpha\rangle = \lim_{\varepsilon\to 0} \frac{i\varepsilon}{E_{pq} + i\varepsilon - \frac{3}{4}q_\alpha^2 - h_\alpha^0 - V_\alpha}|p_\alpha q_\alpha\rangle$$

$$= \lim_{\varepsilon\to 0} \frac{i\varepsilon}{p_\alpha^2 + i\varepsilon - h_\alpha^0 - V_\alpha}|p_\alpha q_\alpha\rangle = |p_\alpha\rangle^{(+)}|q_\alpha\rangle \,. \qquad (3.54)$$

In the second equality we have used the energy relation (3.13) and recognized the Möller wave operator for the pair hamiltonian $h_\alpha^0 + V_\alpha$. Its application to the momentum eigenstate $|p_\alpha\rangle$ yields the two-body scattering state $|p_\alpha\rangle^{(+)}$ as we have seen in Sect. 1.1. Thus $|\Psi_0^{(+)}\rangle$ obeys the inhomogeneous Lippmann-Schwinger equations

$$|\Psi_0^{(+)}\rangle = |p_\alpha\rangle^{(+)}|q_\alpha\rangle + G_\alpha V^\alpha |\Psi_0^{(+)}\rangle, \quad \alpha = 1, 2, 3 \tag{3.55}$$

and can therefore not be admixed in a solution of the set (3.51 a – c).

In contrast to potential scattering, where the Lippmann-Schwinger equation was sufficient to fix the asymptotic behaviour, we need now three Lippmann-Schwinger equations, two of which are homogeneous, to handle the three two-body fragmentation channels. Since $|\Psi_\alpha^{(+)}\rangle$ is uniquely defined, its behaviour in the break-up channel is no longer free and could be read off from each of the equations (3.51). Instead one may use

$$|\Psi_\alpha^{(+)}\rangle = G_0(V + V_4)|\Psi_\alpha^{(+)}\rangle, \tag{3.56}$$

which is also a valid equation [see (3.36) and Lippmann's identity (3.37) for $\beta = 0$]. It guarantees a purely outgoing wave in the break-up channel.

Since the handling of Lippmann-Schwinger equations on the real axis ($\varepsilon = 0$) lends itself very easily to pitfalls [3.7 – 9] we would like to add an example to demonstrate the necessity of being very careful on leaving the safe domain $\varepsilon > 0$. One may ask whether the homogeneous equations (3.51 b, c) are consistent with the inhomogeneous one (3.51 a). Take as an example an arbitrary solution of the homogeneous equation with the kernel $G_2 V^2$:

$$|\Psi\rangle = G_2 V^2 |\Psi\rangle \tag{3.57a}$$

$$= G_1 V^2 |\Psi\rangle + (G_2 - G_1) V^2 |\Psi\rangle \tag{3.57b}$$

$$\overset{?}{=} G_1 V^2 |\Psi\rangle + G_1(V_2 - V_1) G_2 V^2 |\Psi\rangle \tag{3.57c}$$

$$\overset{?}{=} G_1 V^2 |\Psi\rangle + G_1(V_2 - V_1) |\Psi\rangle \tag{3.57d}$$

$$= G_1 V^1 |\Psi\rangle. \tag{3.57e}$$

This is obviously inconsistent with the first equation (3.51 a)

$$|\Psi\rangle = |\phi_1\rangle + G_1 V^1 |\Psi\rangle. \tag{3.58}$$

Where are the traps? The first equality marked with a question mark uses the resolvent identity

$$G_2 - G_1 = G_1(V_2 - V_1) G_2, \tag{3.59}$$

which introduces an additional integration. One has a sequence of operations; in coordinate space a sequence of integrations, the order of which is not specified in this notation. In arriving at the following equality one uses (3.57a) which makes an explicit choice for the order of integration. That is the trap.

It was *Gerjuoy* [3.10] who discussed these type of questions at great length. Thus let us be more careful and put

$$|\Psi\rangle = G_1 V^2 |\Psi\rangle + \{G_1(V_2 - V_1)G_2\} V^2 |\Psi\rangle$$
$$= G_1 V^2 |\Psi\rangle + G_1(V_2 - V_1)\{G_2 V^2 |\Psi\rangle\}$$
$$+ \{G_1(V_2 - V_1)G_2\} V^2 |\Psi\rangle - G_1(V_2 - V_1)\{G_2 V^2 |\Psi\rangle\}, \tag{3.60}$$

where the curly brackets mean that integration should be carried out first inside the brackets. Now using the first equation in (3.57) in the form

$$(E - H_2)|\Psi\rangle = V^2 |\Psi\rangle \tag{3.61}$$

we get

$$|\Psi\rangle = G_1 V^1 |\Psi\rangle + \{G_1(V_2 - V_1)G_2\}(E - H_2)|\Psi\rangle - G_1(V_2 - V_1)|\Psi\rangle$$
$$= G_1 V^1 |\Psi\rangle + \{G_1(V_2 - V_1)G_2\}(E - \overleftarrow{H}_2 + \overrightarrow{H}_2 - \overrightarrow{H}_2)|\Psi\rangle$$
$$- G_1(V_2 - V_1)|\Psi\rangle . \tag{3.62}$$

The arrows on H_2 indicate that the differential operators act either to the left or to the right. Now on the real axis,

$$G_2(E - \overleftarrow{H}_2) = 1 \tag{3.63a}$$

is still valid of course, whereas

$$G_2(E - \overrightarrow{H}_2) \neq 1 \tag{3.63b}$$

in general. Only if the left hand side is applied onto a state which vanishes sufficiently fast at infinity, does a partial integration connect (3.62) with (3.61) without occurrence of surface or correction terms. Thus using (3.59) again, we are left with

$$|\Psi\rangle = G_1 V^1 |\Psi\rangle + (G_1 - G_2)\overrightarrow{H}_2 |\Psi\rangle , \tag{3.64}$$

where $\overrightarrow{H}_2 = \overrightarrow{H}_2 - \overleftarrow{H}_2 = \overrightarrow{H}_0 - \overleftarrow{H}_0 \equiv \overrightarrow{H}_0$. Now

$$G_2 \overrightarrow{H}_2 |\Psi\rangle = G_2(\overrightarrow{H}_2 - E + E - \overleftarrow{H}_2)|\Psi\rangle = |\Psi\rangle - G_2 V^2 |\Psi\rangle = 0 \tag{3.65}$$

and we finally get

$$|\Psi\rangle = G_1 V^1 |\Psi\rangle + G_1 \overrightarrow{H}_0 |\Psi\rangle . \tag{3.66}$$

Clearly the second term is a surface integral in the six-dimensional space of relative motions, and it can be easily evaluated [3.7] with the aid of the asymptotic behaviour of G_1 and $|\Psi\rangle$. As a short-cut let us regard

$$G_1 \vec{H}_0 |\phi_1\rangle = G_1 (\vec{H}_1 - E + E - \bar{H}_1) |\phi_1\rangle = |\phi_1\rangle . \tag{3.67}$$

It turns out, that for $|\Psi\rangle = |\Psi_1\rangle$, only the incoming part which belongs to $|\phi_1\rangle$ contributes to the surface integral and one gets

$$G_1 \vec{H}_0 |\Psi_1\rangle = |\phi_1\rangle . \tag{3.68a}$$

Thus for $|\Psi\rangle = |\Psi_1\rangle$, (3.57a) and (3.58) are indeed compatible. However, the discussion above told us that (3.57a) alone allows the general solution

$$|\Psi\rangle = \alpha |\Psi_1\rangle + \beta |\Psi_3\rangle . \tag{3.69}$$

Therefore the surface integral has to be studied for $|\Psi_3\rangle$ as well, which yields

$$G_1 \vec{H}_0 |\Psi_3\rangle = 0 \tag{3.68b}$$

and (3.57a, 58) are generally compatible. In addition, we see again that (3.57a) and (3.58) do not define $|\Psi\rangle$ uniquely, since $|\Psi_3\rangle$ may be still admixed. It is exactly the second homogeneous equation in (3.51) which has to be added to exclude $|\Psi_3\rangle$.

The considerations of this section can be easily generalised to four and more particles [3.11, 12]. We shall discuss the case of four particles in Chap. 4. For four particles, the number of two-body fragmentation channels is seven, and correspondingly seven Lippmann-Schwinger equations are required to specify the scattering state uniquely.

3.3 Faddeev Equations and Other Coupling Schemes

3.3.1 Faddeev Equations

The set (3.51 a – c) defines the scattering state $|\Psi_\alpha^{(+)}\rangle$ uniquely ($\alpha = 1, 2,$ or 3). Similarily in the case of $\alpha = 0$, the set of three equations (3.55) is necessary and sufficient to define $|\Psi_0^{(+)}\rangle$ uniquely. In each case the same state has to fulfill three different equations. Through these sets do not provide directly a practical algorithm it is a small step from that basis to achieve coupled equations. To simplify matters let us first neglect the possible presence of the three-body force V_4. Then the form (3.56) suggests a decomposition of the total state into 3 parts:

$$|\Psi_\alpha^{(+)}\rangle = \sum_{\mu=1}^{3} G_0 V_\mu |\Psi_\alpha^{(+)}\rangle \equiv \sum_{\mu=1}^{3} |\psi_{\alpha,\mu}\rangle , \qquad (3.70)$$

where

$$|\psi_{\alpha,\mu}\rangle \equiv G_0 V_\mu |\Psi_\alpha^{(+)}\rangle . \qquad (3.70\,a)$$

Operate now by $G_0 V_\alpha$, $G_0 V_\beta$ and $G_0 V_\gamma$ from the left on $(3.51\,a-c)$, respectively. Equation $(3.51\,a)$ for instance turns into

$$|\psi_{\alpha,\alpha}\rangle = G_0 V_\alpha |\phi_\alpha\rangle + G_0 V_\alpha G_\alpha (V_\beta + V_\gamma) |\Psi_\alpha^{(+)}\rangle . \qquad (3.71)$$

Since $G_0 V_\alpha G_\alpha = G_\alpha V_\alpha G_0$, the components $|\psi_{\alpha,\beta}\rangle$ and $|\psi_{\alpha,\gamma}\rangle$ show up again on the right hand side. Furthermore the driving term is just $|\phi_\alpha\rangle$ as we pointed out in (3.50). Therefore we get altogether

$$
\begin{aligned}
|\psi_{\alpha,\alpha}\rangle &= |\phi_\alpha\rangle + G_\alpha V_\alpha (|\psi_{\alpha,\beta}\rangle + |\psi_{\alpha,\gamma}\rangle) \\
|\psi_{\alpha,\beta}\rangle &= \qquad\quad G_\beta V_\beta (|\psi_{\alpha,\gamma}\rangle + |\psi_{\alpha,\alpha}\rangle) \\
|\psi_{\alpha,\gamma}\rangle &= \qquad\quad G_\gamma V_\gamma (|\psi_{\alpha,\alpha}\rangle + |\psi_{\alpha,\beta}\rangle) .
\end{aligned}
\qquad (3.72)
$$

This is a set of three coupled equations for the amplitudes $|\psi_{\alpha,\varepsilon}\rangle$, $\varepsilon = 1, 2, 3$, which according to (3.70) sum up to the total state $|\Psi_\alpha^{(+)}\rangle$. They are the Faddeev equations [3.5].

The importance of this set and an additional insight into the structure of few body equations, justify that we present two further derivations.

To simplify the notation, let us drop for the present the index α indicating the initial channel. The decomposition (3.70) tells us that the components obey the following set of three coupled equations:

$$|\psi_\alpha\rangle \equiv G_0 V_\alpha |\Psi\rangle = G_0 V_\alpha \sum_\beta |\psi_\beta\rangle . \qquad (3.73)$$

Let us iterate this set once:

$$|\psi_\alpha\rangle = G_0 V_\alpha \sum_\beta G_0 V_\beta \sum_\gamma |\psi_\gamma\rangle . \qquad (3.74)$$

We recognize that, if we would continue to iterate, there would result different types of operator sequences. One of them is singled out in that *only* V_α acts between free propagators G_0. In all other sequences at least one other additional pair interaction is present. In this special sequence, $G_0 V_\alpha G_0 V_\alpha \ldots$, the particle α does not feel an interaction, and one faces only a two-body problem which can be solved formally. To that end we separate the term $\beta = \alpha$ on the rhs of (3.73) and shift it to the lhs:

$$(1 - G_0 V_\alpha) |\psi_\alpha\rangle = G_0 V_\alpha \sum_{\beta \neq \alpha} |\psi_\beta\rangle \; . \tag{3.75}$$

Using now the operator $(1 + G_\alpha V_\alpha)$ we can solve for $|\psi_\alpha\rangle$:

$$|\psi_\alpha\rangle = |\overset{0}{\psi_\alpha}\rangle + (1 + G_\alpha V_\alpha) G_0 V_\alpha \sum_{\beta \neq \alpha} |\psi_\beta\rangle \; . \tag{3.76}$$

This is the general solution of (3.75) including the driving term, which is a solution of the lhs alone. We saw in the step from (3.71) to (3.72) that this is the channel state $|\phi_\alpha\rangle$. Furthermore, because of the resolvent identity between G_α and G_0, the kernel simplifies and we end up with

$$|\psi_\alpha\rangle = |\phi_\alpha\rangle + G_\alpha V_\alpha \sum_{\beta \neq \alpha} |\psi_\beta\rangle \quad \alpha, \beta = 1, 2, 3 \; . \tag{3.77}$$

This set of three coupled equations must have now the property that iterations cannot yield an operator sequence with only V_α acting. This is obviously the case.

The general solution (3.77), however, is not the physical one, which is defined through the set (3.72). In (3.77) there are nonzero driving terms in all three equations. This is not surprising since we started from the homogeneous Lippmann-Schwinger equation (3.70), where the boundary conditions are not specified, while (3.72) was derived from the set (3.51), which had the physical boundary conditions built in. We shall see in the next section that the second terms on the right hand side of (3.77) are asymptotically purely outgoing. Consequently $|\phi_\alpha\rangle$ can be kept as a driving term in (3.77) only in the initial channel, and we are back at (3.72).

The third derivation emphasizes the insight into connected and disconnected structures. It is close to the one presented by *Faddeev* in his classic paper [3.5]. Let us regard the resolvent identity

$$G = G_0 + G_0 \sum_\mu V_\mu G \; , \tag{3.78}$$

which is an integral equation for the full resolvent operator G. The kernel, $G_0 \sum_\mu V_\mu$, consists of three parts, in each of which only one pair interacts. Therefore in momentum representation $\langle k_1' k_2' k_3' | G_0 \sum_\mu V_\mu | k_1 k_2 k_3 \rangle$, the noninteracting particle cannot undergo a momentum change and we encounter three different δ-functions in the sum. Because of that, the kernel cannot be of the Fredholm or Hilbert-Schmidt type, which would allow standard approximations in numerical approaches. Can one rewrite this integral equation in another form which belongs to the Fredholm family? One can get insight into the mechanism of (3.78) by expanding the rhs into its Neumann series

$$G = G_0 + G_0 \sum_\mu V_\mu G_0 + G_0 \sum_\mu V_\mu G_0 \sum_\nu V_\nu G_0 + \cdots \; . \tag{3.79}$$

It is very useful to present this series graphically. We introduce

$$G_0 \triangleq \equiv$$

$$V_\mu \triangleq \}$$

(3.80)

Then (3.79) is

(3.81)

Clearly there are three infinite subseries of diagrams like

where one particle does not interact. These are called disconnected diagrams and are responsible for the δ-functions in momentum space mentioned above. In the remaining diagrams all particles interact and they are called connected. Each of the infinite subseries of disconnected diagrams is known to us. Together with G_0 it is just the channel resolvent operator. For instance

(3.82)

Therefore we can reorder (3.81) as

(3.83)

Note that the third term collects all diagrams of the type

Through the use of the channel resolvent operator G_α one sums up all the infinite subseries where only two particles interact. The next step is to decompose G into three parts according to the leftmost interaction:

with   (3.84)

(3.85)

Obviously the structures following $G_\alpha V_\alpha \equiv$ to the right all have either V_β or V_γ as the leftmost interaction. Therefore they are the parts $G^{(\beta)}$ and $G^{(\gamma)}$ of G. Thus we end up with

(3.86)

(Of course α, β, and γ are all different).

This is a set of three coupled equations linking the three sets of diagrams (3.85). Starting from the definition (3.84) the set (3.86) can easily be determined algebraically as well.

Exercise: Derive (3.86) algebraically.

What are the consequences now for the scattering states $|\Psi_{\alpha_0}^{(+)}\rangle$ which are defined by (3.24, 25). According to the decomposition (3.84) and dropping again the index α_0 we have

$$|\Psi^{(+)}\rangle = \lim_{\varepsilon \to 0} i\varepsilon G_0 |\phi_{\alpha_0}\rangle + \lim_{\varepsilon \to 0} i\varepsilon \sum_\alpha G^{(\alpha)} |\phi_{\alpha_0}\rangle .$$ (3.87)

The first term vanishes for $\alpha_0 \neq 0$ according to (3.37) and the second one defines the three amplitudes

$$|\psi_\alpha\rangle \equiv \lim_{\varepsilon \to 0} i\varepsilon G^{(\alpha)} |\phi_{\alpha_0}\rangle .$$ (3.88)

Using now the coupled set (3.86) we find immediately

$$|\psi_\alpha\rangle = |\phi_{\alpha_0}\rangle \delta_{\alpha\alpha_0} + \sum_{\beta\neq\alpha} G_\alpha V_\alpha |\psi_\beta\rangle . \tag{3.89}$$

Faddeev [3.13] proved that this set of equations has a unique solution. We shall use the direct link to the unique set of Lippmann-Schwinger equations, which we regarded in the beginning of this section, to proof the uniqueness of the set (3.89). Assume two sets of amplitudes would solve the set (3.89). Then the difference, $|\chi_\alpha\rangle$, would obey the homogeneous set

$$|\chi_\alpha\rangle = G_\alpha V_\alpha \sum_{\beta\neq\alpha} |\chi_\beta\rangle . \tag{3.90}$$

This can be rewritten as

$$|\chi_\alpha\rangle = G_0 V_\alpha (1 + G_\alpha V_\alpha) \sum_{\beta\neq\alpha} |\chi_\beta\rangle , \tag{3.91}$$

which defines an auxiliary amplitude $|\theta_\alpha\rangle$:

$$|\chi_\alpha\rangle \equiv G_0 V_\alpha |\theta_\alpha\rangle . \tag{3.92}$$

However $|\theta_\alpha\rangle$ does not depend on α because

$$|\theta_\alpha\rangle = (1 + G_\alpha V_\alpha) \sum_{\beta\neq\alpha} |\chi_\beta\rangle = \sum_{\beta\neq\alpha} |\chi_\beta\rangle + |\chi_\alpha\rangle$$

$$= \sum_\gamma |\chi_\gamma\rangle \equiv |\Theta\rangle . \tag{3.93}$$

Therefore (3.91) is equivalent to

$$G_0 V_\alpha \left[|\Theta\rangle - (1 + G_\alpha V_\alpha) \sum_{\beta\neq\alpha} G_0 V_\beta |\Theta\rangle \right] = 0 \tag{3.94}$$

or

$$G_0 V_\alpha [|\Theta\rangle - G_\alpha V^\alpha |\Theta\rangle] = 0 ; \quad \alpha = 1, 2, 3 . \tag{3.95}$$

We conclude that the curly bracket should vanish, so that $|\Theta\rangle$ satisfies the unique set of homogeneous Lippmann-Schwinger equations. The only non-trivial solutions to this set are three-body bound states. Thus for energies accessible to scattering $|\chi_\alpha\rangle \equiv 0$ and the Faddeev set (3.89) is unique.

In solving equations of this type, one has to decide which representation is most convenient. We shall consider the representation in configuration space in the next section. One possibility in momentum space is clearly the set of eigenstates (3.9, 12) to H_α which naturally shows up in the spectral representation of G_α. This has been proposed and developed in [3.14] and applied in [3.15]. Another possibility is to work with momentum eigenstates, where it is customary to introduce the two-body t-matrices via

$$G_\alpha V_\alpha \equiv G_0 t_\alpha. \tag{3.96}$$

Using the resolvent identity, we see indeed that t_α obeys the familiar Lippmann-Schwinger equation for the two-body t-matrix

$$t_\alpha = V_\alpha + V_\alpha G_0 t_\alpha. \tag{3.97}$$

Note, however, that t_α is defined now in the three-body space. Thus (3.97) reads explicitely

$$\langle p_\alpha q_\alpha | t_\alpha(z) | p'_\alpha q'_\alpha \rangle = \langle p_\alpha q_\alpha | V_\alpha | p'_\alpha q'_\alpha \rangle$$

$$+ \int dp''_\alpha dq''_\alpha \langle p_\alpha q_\alpha | V_\alpha | p''_\alpha q''_\alpha \rangle \frac{1}{z - p''_\alpha{}^2 - \frac{3}{4} q''_\alpha{}^2}$$

$$\times \langle p''_\alpha q''_\alpha | t_\alpha(z) | p'_\alpha q'_\alpha \rangle. \tag{3.98}$$

Since V_α does not act on the spectator motion we get

$$\langle p_\alpha q_\alpha | t_\alpha(z) | p'_\alpha q'_\alpha \rangle = \delta(q_\alpha - q'_\alpha) \langle p_\alpha | \hat{t}_\alpha(z - \tfrac{3}{4} q_\alpha^2) | p'_\alpha \rangle \tag{3.99}$$

with the genuine two-body t-matrix $\hat{t}_\alpha$ taken at the shifted energy $z - \frac{3}{4} q_\alpha^2$. Thus two-body t-matrices enter into the Faddeev equations as off-shell quantities, as was mentioned in Sect. 1.5. In terms of t_α, the Faddeev equation (3.89) take the standard form

$$\begin{pmatrix} |\psi_1\rangle \\ |\psi_2\rangle \\ |\psi_3\rangle \end{pmatrix} = \begin{pmatrix} |\phi_1\rangle \\ 0 \\ 0 \end{pmatrix} + G_0 \begin{pmatrix} 0 & t_1 & t_1 \\ t_2 & 0 & t_2 \\ t_3 & t_3 & 0 \end{pmatrix} \begin{pmatrix} |\psi_1\rangle \\ |\psi_2\rangle \\ |\psi_3\rangle \end{pmatrix}. \tag{3.100}$$

This set defines $|\Psi_1^{(+)}\rangle$. It should be clear how the driving term is modified for the other scattering states.

Exercise: What is the driving term for $|\Psi_0^{(+)}\rangle$?

Have we now arrived in the safe harbour of the Fredholm family of integral equations? Certainly each element in the Faddeev matrix kernel, K_F, occurring in (3.100) is still disconnected and K_F as a whole is not of the Hilbert-Schmidt type, since $K_F K_F^+$ still contains squares of δ-functions. However in contrast to kernels of Lippmann-Schwinger equations the once iterated Faddeev kernel, K_F^2, is connected. This is obvious because of the zeros in the main diagonal. It is not too difficult to see [3.16, 17] that K_F^2 is of the Hilbert-Schmidt type for $\mathrm{Im}\{z\} \neq 0$ and for well behaved potentials, for instance finite range potentials. So we can answer the above question positively.

3.3.2 Faddeev Equations in Differential Form and the Asymptotic Behaviour of the Faddeev Amplitudes

The asymptotic behaviour of a three-body wave function in coordinate space is richer in structure than for two-particle potential scattering. Since the Faddeev equations have a unique solution, they have to provide that information. A quick glance at (3.72) tells us that the extraction of this information cannot be as straightforward as in (1.81) for potential scattering, where the integration variables x' are confined for instance by an interaction of finite range and therefore the asymptotic limit $|x| \rightarrow \infty$ can be easily taken. For three particles there occur *two* relative vectors, r' and R', as integration variables, and clearly the pair interaction in (3.72) cannot confine both. Thus the Faddeev components, the unknown's themselves, have to take over part of the job of making the integral convergent. It is this conspiracy of the pair interaction V_α in the Faddeev kernel together with the Faddeev components $|\psi_\beta\rangle$, $\beta \neq \alpha$ which is interesting enough to justify the following, rather detailed, discussion.

To avoid all inessential burdens we consider three identical spinless bosons interacting by pure s-wave interactions in the state of total angular momentum $L = 0$. Before embarking let us make a few general remarks. The three sets of Jacobi coordinates (3.3a) are linearly related to each other. We define coordinate states

$$|r_1 R_1\rangle_1 = |r_2 R_2\rangle_2 = |r_3 R_3\rangle_3 , \tag{3.101}$$

where for instance the subscript 1 on the ket symbol tells us that r_1 is the relative distance vector between particles 2 and 3 and R_1 the relative distance vector between particle 1 and the center-of-mass of particles 2 and 3, as described in (3.3a). That same spatial configuration can also be described by the other two types of coordinate states with subscripts 2 and 3. The coordinate representation of the state $|\Psi\rangle$ can therefore be defined through

$$\Psi(r_1 R_1) \equiv {}_1\langle r_1 R_1 | \Psi\rangle = {}_2\langle r_2 R_2 | \Psi\rangle = {}_3\langle r_3 R_3 | \Psi\rangle . \tag{3.102a}$$

Now for three bosons $|\Psi\rangle$ must be symmetric under all permutations of the particles. Therefore the wave function has the property

$$\Psi(r_1 R_1) = {}_1\langle r_1 R_1 | \Psi\rangle = {}_1\langle r_1 R_1 | P_{12} P_{23} | \Psi\rangle$$

$$= {}_3\langle r_1 R_1 | \Psi\rangle = {}_1\langle r_3 R_3 | \Psi\rangle = \Psi(r_3 R_3) \tag{3.102b}$$

and similarily using $P_{13}P_{23}$

$$\Psi(r_1 R_1) = \Psi(r_2 R_2) . \tag{3.102c}$$

Remembering now the definition of the Faddeev components

$$|\psi_\alpha\rangle = G_0 V_\alpha |\Psi\rangle \tag{3.103}$$

we find immediately the result that

$$\psi_2(r_2 R_2) \equiv {}_2\langle r_2 R_2 | \psi_2 \rangle = {}_2\langle r_2 R_2 | P_{12} P_{23} | \psi_1 \rangle$$
$$= {}_1\langle r_2 R_2 | \psi_1 \rangle \equiv \psi_1(r_2 R_2) \tag{3.104a}$$

and similarily

$$\psi_3(r_3 R_3) = \psi_1(r_3 R_3) . \tag{3.104b}$$

In other words, the three Faddeev components have the same functional form if expressed in the "natural" Jacobi coordinates. The three Faddeev equations then reduce to one:

$$\psi_1(r_1 R_1) = \phi_1(r_1 R_1) + \int dr_1' \, dR_1' \, \langle r_1 R_1 | G_1 | r_1' R_1' \rangle$$
$$\times V_1(r_1') [\psi_1(r_2' R_2') + \psi_1(r_3' R_3')] . \tag{3.105}$$

Let us introduce the simplifications. We consider the state of total angular momentum $L = 0$ and assume that V_1 acts only in the s-wave, where it supports one bound state $\varphi_1(r_1)$. The partial-wave decomposition of $\psi_1(r_1, R_1)$ therefore reduces to s-waves in both relative motions, and thus to a dependence on the magnitudes r_1 and R_1 only. As a consequence, the unknown amplitudes under the integral depend only on

$$r_2' = |\tfrac{1}{2}r_1' + R_1'| \qquad r_3' = |\tfrac{1}{2}r_1' - R_1'|$$
$$R_2' = |\tfrac{3}{4}r_1' - \tfrac{1}{2}R_1'| \qquad R_3' = |\tfrac{3}{4}r_1' + \tfrac{1}{2}R_1'|, \tag{3.106}$$

[see (3.3a)] which involve only the single angle between the vectors r_1' and R_1'. It should be obvious that (3.105) reduces to

$$u(r_1, R_1) = u_b(r_1) \sin q_0 R_1 + \int_0^\infty dr_1' \int_0^\infty dR_1'$$
$$\times \left\langle r_1 R_1 \left| \frac{1}{E + i0 - H_0 - V_1} \right| r_1' R_1' \right\rangle V(r_1') Q(r_1' R_1') . \tag{3.107}$$

We defined $u(r_1 R_1) = r_1 R_1 \psi(r_1 R_1)$, $u_b(r_1) = r_1 \varphi(r_1)$ and the source term

$$Q(r_1 R_1) = \int_{-1}^{1} dx \frac{r_1 R_1}{r_2 R_2} u(r_2 R_2) . \qquad (3.108)$$

Now we see explicitely, as remarked above, that the pair interaction $V(r_1')$ in the kernel confines only one variable so that the convergence in R_1' has to be provided by the source term itself. What do we have to know about the unknown Q? Let us assume again an interaction of finite range, r_0. Thus we need to know the behaviour of Q for $r_1 \leq r_0$, $R_1 \to \infty$ in order to prove the convergence of the integral in (3.107). Because of (3.106), however, this is equivalent to r_2 and R_2 going to infinity. The behaviour of u in this limit is elementary and we can deduce it directly from the differential form of the Faddeev equations. Operating with $(E - H_1)$ onto (3.105), one gets

$$(E - H_1) \psi_1(r_1 R_1) = - V_1(r_1) [\psi_1(r_2, R_2) + \psi_1(r_3, R_3)] \qquad (3.109\,a)$$

or in our simplified case

$$\left[-\frac{d^2}{dr_1^2} - \frac{3}{4} \frac{d^2}{dR_1^2} + V_1(r_1) - E \right] u(r_1 R_1) = - V_1(r_1) Q(r_1 R_1) . \qquad (3.109\,b)$$

This is a partial integrodifferential equation, which is an important practical tool for solving the three body problem as we shall point out below. Now we use it just to extract the behaviour of u for $r_1 \to \infty$ and $R_1 \to \infty$. In this limit, (3.109b) reduces to

$$\left(-\frac{d^2}{dr_1^2} - \frac{3}{4} \frac{d^2}{dR_1^2} - E \right) u(r_1 R_1) = 0 . \qquad (3.110)$$

It is convenient to introduce polar coordinates

$$\begin{aligned} r_1 &= \rho \cos \varphi \\ R_1 &= \sqrt{\tfrac{3}{4}} \rho \sin \varphi . \end{aligned} \qquad (3.111)$$

Then one easily finds two linearly independent solutions of (3.110) in the limit $\rho \to \infty$:

$$u(\rho, \varphi) \to \frac{e^{\pm i\sqrt{E}\rho}}{\rho^{1/2}} A(\varphi) . \qquad (3.112)$$

What is needed is only the $\rho^{-1/2}$ dependence. We conclude that for r_1 fixed and $R_1 \to \infty$ the source term behaves as

$$Q(r_1 R_1) \to O\left(\frac{1}{R_1^{3/2}}\right). \tag{3.113}$$

This behaviour is the essential key and guarantees absolute convergence of the integral over R_1' in (3.107). It is based on purely geometrical aspects: the asymptotic form (3.112) in two dimensions and the distance behaviour of Jacobi coordinates for rearrangement channels. Of course these geometrical properties can be effective in this manner only due to the basic structure of the Faddeev equations, which link only rearrangement channels.

We can now proceed and regard the spectral representation of the Green's function in (3.107). The completeness relation, restricted to s-states only, reads

$$\frac{2}{\pi}\int_0^\infty dq \, \sin(qR_1)\sin(qR_1')u_b(r_1)u_b(r_1')$$

$$+ \left(\frac{2}{\pi}\right)^2 \int_0^\infty dp \int_0^\infty dq \, \sin(qR_1)\sin(qR_1')u_p^{(-)}(r_1)u_p^{(-)}(r_1')^*$$

$$= \delta(r_1-r_1')\,\delta(R_1-R_1'), \tag{3.114}$$

where two-body s-wave scattering states are normalised as

$$u_p^{(\pm)}(r) \xrightarrow[r\to\infty]{} e^{\pm i\delta(p)}\sin[pr+\delta(p)]. \tag{3.115}$$

This leads to the Green's function

$$\left\langle r_1 R_1 \left| \frac{1}{E+i0-H_0-V_1} \right| r_1' R_1' \right\rangle$$

$$= u_b(r_1)\frac{2}{\pi}\int_0^\infty dq \, \frac{\sin qR_1 \sin qR_1'}{E+i0-\varepsilon_1-\frac{3}{4}q^2}u_b(r_1')$$

$$+ \left(\frac{2}{\pi}\right)^2 \int_0^\infty dp \int_0^\infty dq \, \frac{\sin(qR_1)u_p^{(-)}(r_1)\sin(qR_1')u_p^{(-)*}(r_1')}{E+i0-p^2-\frac{3}{4}q^2}. \tag{3.116}$$

The integral in q defines a free Green's function and can be carried through using elementary function theory. We obtain

$$\langle r_1 R_1 | G_1 | r_1' R_1'\rangle = u_b(r_1)\left(-\frac{4}{3}e^{iq_0 R_>}\frac{\sin q_0 R_<}{q_0}\right)u_b(r_1')$$

$$+ \frac{2}{\pi}\int_0^\infty dp \, u_p^{(-)}(r_1)\left(-\frac{4}{3}e^{iq_p R_>}\frac{\sin q_p R_<}{q_p}\right)u_p^{(-)*}(r_1'), \tag{3.117}$$

where the wave numbers are

$$q_0^2 = \tfrac{4}{3}(E - \varepsilon_1)$$
$$q_p^2 = \tfrac{4}{3}(E - p^2)$$

(3.118)

and $R_\langle = \min(R_1, R_1')$, $R_\rangle = \max(R_1, R_1')$.

What asymptotic behaviour of u do we expect? The driving term in (3.107) provides an ingoing and outgoing unperturbed radial wave for particle 1 with respect to the bound pair. The interaction should lead to scattered outgoing waves in all 4 channels. Applying the first term on the rhs of (3.117) to VQ yields

$$-\frac{4}{3} e^{iq_0 R_1} u_b(r_1) \int_0^{R_1} dR_1' \frac{\sin q_0 R_1'}{q_0} \int_0^\infty dr_1' u_b(r_1') V_1(r_1') Q(r_1' R_1')$$

$$-\frac{4}{3} \frac{\sin q_0 R_1}{q_0} u_b(r_1) \int_0^\infty dR_1' e^{iq_0 R_1'} \int_0^\infty dr_1' u_b(r_1') V_1(r_1') Q(r_1' R_1').$$

(3.119)

In the limit $R_1 \to \infty$ this tends towards

$$-\tfrac{4}{3} u_b(r_1) e^{iq_0 R_1} T_b$$

(3.120)

with

$$T_b = \int_0^\infty dR \int_0^\infty dr \frac{\sin q_0 R}{q_0} u_b(r) V(r) Q(rR).$$

(3.121)

As expected we find an outgoing scattered wave in channel 1 and the amplitude T_b has the correct form (3.31). Indeed, inserting the definition (3.103) of the Faddeev amplitudes, we get

$$T_b = \langle \phi_1 | V_1 | \psi_2 + \psi_3 \rangle = \langle \phi_1 | V_1 G_0 (V_2 + V_3) | \Psi^{(+)} \rangle$$

$$= \langle \phi_1 | V_2 + V_3 | \Psi^{(+)} \rangle.$$

(3.122)

In (3.122) we have switched back to a three-dimensional notation and used in the last equality (3.50). Thus we find as the first result, that the asymptotic behaviour of the wave function in channel 1 its contained completely in the Faddeev component $| \psi_1 \rangle$, that of channel 2 in $| \psi_2 \rangle$ and that of channel 3 in $| \psi_3 \rangle$. It remains to pin down the asymptotic form in the break-up channel. Since asymptotically the three particles cannot interact anymore, the two energies of relative motion have to add up to the total energy E. Therefore to the asymptotic flux, only part of the second term in (3.117) can contribute, and we find

$$-\frac{4}{3}\frac{2}{\pi}\int_0^{\sqrt{E}} dp\, u_p^{(-)}(r_1)\, e^{iq_p R_1} T(p)$$

(3.123)

with

$$T(p) = \int_0^\infty dR \int_0^\infty dr\, \frac{\sin q_p R}{q_p}\, u_p^{(-)*}(r)\, V(r)\, Q(rR)\,.$$

(3.124)

This is discussed in [3.18] and the remaining terms are shown to be correction terms of the asymptotic form of u, which may be interesting for a numerical analysis, but do not contribute to the asymptotic flux of particles. The expression (3.123) however is not yet the final form, since the integrand oscillates in the asymptotic limits of r_1 and R_1 and the leading term for the integral has to be extracted. This can be done by the method of steepest descent [3.19] and is described in [3.18]. What do we expect? The energy can be partitioned continuously among the two relative motions. Corresponding to every situation there is a certain ratio r_1/R_1, which in polar coordinates fixes the angle φ. This should be reflected in the argument of T in the asymptotic form. One finds

$$u(r_1 R_1) \xrightarrow[\rho \to \infty]{} \frac{4}{3}\sqrt{\frac{2}{\pi}}\, e^{i\pi/4} E^{1/4} e^{i\sqrt{E}\rho}/\rho^{1/2} \sin\varphi\, T(\sqrt{E}\cos\varphi)\,.$$

(3.125)

Again we can rewrite $T(p)$ back in a three-dimensional notation as

$$\frac{T(p)}{p} = \langle \phi_p^{(-)}|V_1|\psi_2 + \psi_3 \rangle\,.$$

(3.126)

Since the state $\langle \phi_p^{(-)}|$ can be represented as

$$\langle \phi_p^{(-)}| = \langle pq| + \langle pq|V_1 G_1$$

(3.127)

the amplitude $T(p)/p$ is

$$\begin{aligned}
\frac{T(p)}{p} &= \langle pq|V_1|\psi_2 + \psi_3 \rangle + \langle pq|V_1 G_1 V_1(\psi_2 + \psi_3)\rangle \\
&= \langle pq|V_1|(\psi_2 + \psi_3)\rangle + \langle pq|V_1|(\psi_1 - \phi_1)\rangle \\
&= \langle pq|V_1|\Psi\rangle - \langle pq|V_1|\phi_1\rangle\,.
\end{aligned}$$

(3.128)

The second term in the last line vanishes on shell (the infinitely strong surface oscillations resulting from partial integration do not contribute to the cross section, which includes an integration over at least a small momentum interval). The first term however is part of the break-up amplitude as given in (3.31) for $\beta = 0$.

The total break-up behaviour adds up coherently from the three Faddeev components. Let us consider a definite break-up configuration described by r_1, R_1 and therefore also by related $r_2 R_2$ and $r_3 R_3$. The lengths r_1, R_1 are equivalent to a certain choice of ρ and φ_1. Similarily the angles φ_2 and φ_3 are linked to $r_2 R_2$ and $r_3 R_3$, respectively. Thus defining $p_i = \sqrt{E} \cos \varphi_i$ and using (3.70), (3.104) and (3.125) we get

$$
\Psi \xrightarrow[\substack{r_i \to \infty \\ R_i \to \infty}]{} e^{i\pi/4} \sqrt{\frac{2}{\pi}} \left(\frac{4}{3}\right)^{3/2} E^{3/4} e^{i\sqrt{E}\rho}/\rho^{5/2}
$$

$$
\times \left[\frac{T(p_1)}{p_1} + \frac{T(p_2)}{p_2} + \frac{T(p_3)}{p_3} \right]. \tag{3.129}
$$

Note that the six-dimensional volume element in polar coordinates is proportional to ρ^5 and therefore the $\rho^{-5/2}$ behaviour in the wave function is necessary to guarantee constant flux through various shells of hyperradius ρ.

We remark that a second form of G_1 can also be used to study the asymptotic behaviour of u (see [3.18]).

The asymptotic form of the Faddeev amplitudes and the total state have been discussed also in [3.20]. The Grenoble group pioneered very successfully the direct solution of the partial differential equations of the type (3.109) for bound [3.21] and scattering states [3.20].

More recent work on that line can be found in a continuing sequence of papers by the Los Alamos group [3.22].

3.3.3 Other Coupling Schemes and Spuriosities

The fundamental set of Lippmann-Schwinger equations can be turned into a coupled set of equations in other manners besides the Faddeev one. Consider for instance the first equation for the state $|\Psi\rangle \equiv |\Psi_1^{(+)}\rangle$:

$$
|\Psi\rangle = |\phi\rangle + G_1(V_2|\Psi\rangle + V_3|\Psi\rangle). \tag{3.130}
$$

In nuclear physics, the pair interactions are of short range and it is a good approximation to assume that they act only in a certain number of low partial waves. To implement that concept it is necessary to develop $|\Psi\rangle$ into angular momentum states $\mathcal{Y}_{l\lambda}^{LM}(\hat{r}\hat{R})$ for the two relative motions. Here l and λ are the orbital angular momenta related to $\hat{r}$ and $\hat{R}$, respectively, and L is the total orbital angular momentum. Obviously we need all three types of Jacobi coordinates to deal with V_1, V_2, and V_3. Thus $|\Psi\rangle$ in $V_2 |\Psi\rangle$ should be expanded with respect to l_2, λ_2 and $|\Psi\rangle$ in $V_3 |\Psi\rangle$ in terms of l_3, λ_3. Let us introduce the projection operators onto a certain number of low partial-wave states:

$$P_\alpha \equiv \sum_{l_\alpha\lambda_\alpha}^{\text{finite}} \mathcal{Y}_{l_\alpha\lambda_\alpha}^{LM}(\hat{r}_\alpha\hat{R}_\alpha)\, \mathcal{Y}_{l_\alpha\lambda_\alpha}^{LM*}(\hat{r}'_\alpha\hat{R}'_\alpha)\,. \tag{3.131}$$

Then (3.130) is approximated by

$$|\Psi\rangle \approx |\phi\rangle + G_1 V_2 P_2 |\Psi\rangle + G_1 V_3 P_3 |\Psi\rangle\,. \tag{3.132}$$

Of course for model interactions, which act by definition only in a finite number of partial waves and which are very common for few body nuclear models, (3.132) is exact. In the two homogeneous Lippmann-Schwinger equations $P_1|\Psi\rangle$ shows up in addition. Therefore we get a closed set if we project the three equations onto P_α, respectively. This results in the coupled set [3.6]

$$\begin{aligned}
P_1|\Psi\rangle &\approx P_1|\phi\rangle + P_1 G_1 V_2 P_2 |\Psi\rangle + P_1 G_1 V_3 P_3 |\Psi\rangle \\
P_2|\Psi\rangle &\approx \qquad\qquad P_2 G_2 V_3 P_3 |\Psi\rangle + P_2 G_2 V_1 P_1 |\Psi\rangle \\
P_3|\Psi\rangle &\approx \qquad\qquad P_3 G_3 V_1 P_1 |\Psi\rangle + P_3 G_3 V_2 P_2 |\Psi\rangle\,.
\end{aligned} \tag{3.133}$$

This set has an interesting property, namely that each kernel $P_\alpha G_\alpha V_\beta$, $\beta \neq \alpha$, is already of the Hilbert-Schmidt type [3.23]. The reason is of purely kinematical origin. The truncated, partial-wave projection makes the two body forces appear like three-body ones in the Hilbert-Schmidt norm. Consider as an example $V(r) = \lambda\theta(a-r)$ and let P project onto s-waves in both relative orbital momenta. Then calculating the norm for $P_1 G_1 V_2$ one encounters

$$P_1 V_2^2(|r_2|) = P_1 V_2^2(|\tfrac{1}{2}r_1 - R_1|) P_1 \tag{3.134}$$

$$\propto \frac{1}{r_1 R_1}\,\theta(a - |\tfrac{1}{2}r_1 - R_1|)\{[\min(a, \tfrac{1}{2}r_1 + R_1)]^2 - (\tfrac{1}{2}r_1 - R_1)^2\}\,,$$

which vanishes not only when $r_2 \to \infty$ but also if r_2 remains small and $R_2 \to \infty$. Finite rank approximations based on this Hilbert-Schmidt property have been proposed in a couple of papers and applied to a three-body bound state model [3.24].

The coupled set (3.133) may tempt us to introduce in (3.51) three unknown functions $|\Psi^{(\alpha)}\rangle$ analogous to $P_\alpha|\Psi\rangle$ and to rewrite it as

$$\begin{aligned}
|\Psi^{(1)}\rangle &= |\phi\rangle + G_1 V_2 |\Psi^{(2)}\rangle + G_1 V_3 |\Psi^{(3)}\rangle \\
|\Psi^{(2)}\rangle &= \qquad\quad G_2 V_3 |\Psi^{(3)}\rangle + G_2 V_1 |\Psi^{(1)}\rangle \\
|\Psi^{(3)}\rangle &= \qquad\quad G_3 V_1 |\Psi^{(1)}\rangle + G_3 V_2 |\Psi^{(2)}\rangle\,.
\end{aligned} \tag{3.135}$$

This form has been proposed for identical particles also in [3.6] (see Sect. 3.4.4) and in another context in [3.25]. From our point of view, the superscripts appear only as dummy variables, since they are meant to define the coupling scheme but not to distinguish three different states. One expects of course that for all α $|\Psi^{(\alpha)}\rangle \equiv |\Psi\rangle$ is the physical solution. Indeed one can prove this in the following manner [3.11]. We operate on the first equation with $(1 - G_0 V_1)$ from the left. Then we get

$$|\Psi^{(1)}\rangle = \sum_{\mu=1}^{3} G_0 V_\mu |\Psi^{(\mu)}\rangle . \tag{3.136}$$

In the same manner, the second and third equations yield $|\Psi^{(2)}\rangle$ and $|\Psi^{(3)}\rangle$ to be equal to the same expression as found for $|\Psi^{(1)}\rangle$ after operating from the left by $(1 - G_0 V_2)$ and $(1 - G_0 V_3)$, respectively. Therefore $|\Psi^{(i)}\rangle$ is indeed independent of the dummy index i. Note that the quantities $|\Psi^{(i)}\rangle$ are not parts of the total wave function as are the Faddeev amplitudes, but stand for the total wave function. It should also be emphasized that the set (3.135) defines a kernel whose square is already connected. This was the main motivation in proposing this system [3.25].

Another coupling scheme was introduced in [3.25] as well. It relies again on dummy variables and we present the following example:

$$|\Psi^{(1)}\rangle = \phi + G_1 V^1 |\Psi^{(2)}\rangle$$
$$|\Psi^{(2)}\rangle = \quad\quad G_2 V^2 |\Psi^{(3)}\rangle \tag{3.137}$$
$$|\Psi^{(3)}\rangle = \quad\quad G_3 V^3 |\Psi^{(1)}\rangle .$$

Introducing the column vector $\Psi^T = (|\Psi^{(1)}\rangle, |\Psi^{(2)}\rangle, |\Psi^{(3)}\rangle)$, the diagonal matrices $(\underset{\approx}{G})_{ij} = \delta_{ij} G_i$ and $(\underset{\approx}{V})_{ij} = \delta_{ij} V^j$ and the so called channel array coupling matrix

$$\underset{\approx}{W} = \begin{pmatrix} 0 & 1 & 0 \\ 0 & 0 & 1 \\ 1 & 0 & 0 \end{pmatrix} , \tag{3.138}$$

the set (3.137) reads in matrix notation

$$\Psi = \phi + \underset{\approx}{G} \underset{\approx}{V} \underset{\approx}{W} \Psi . \tag{3.139}$$

Why such a coupling scheme? One starts from individual kernels $G_i V^i$ which are not connected. A first iteration of (3.139) yields kernels $G_i V^i G_j V^j (i \neq j)$ which still contain disconnected parts. However, after iterating a second time, the kernel becomes fully connected.

Again one has to ask whether this set has *only* the physical solution $|\Psi^{(i)}\rangle = |\Psi\rangle$? Since both sets (3.135) and (3.137) have the solution $|\Psi^{(i)}\rangle = |\Psi\rangle$ one can expect the connection

$$\begin{pmatrix} 1 & -G_1 V^1 & 0 \\ 0 & 1 & -G_2 V^2 \\ -G_3 V^3 & 0 & 1 \end{pmatrix} = M \begin{pmatrix} 1 & -G_1 V_2 & -G_1 V_3 \\ -G_2 V_1 & 1 & -G_2 V_3 \\ -G_3 V_1 & -G_3 V_2 & 1 \end{pmatrix}.$$

$$(3.140)$$

Indeed, it is a relatively easy exercise to work out M [3.26]:

$$\underset{\sim}{M} = \begin{pmatrix} 1 & -G_1 V_3 G_0 G_2^{-1} & G_1 V_3 G_0 G_3^{-1} \\ G_2 V_1 G_0 G_1^{-1} & 1 & -G_2 V_1 G_0 G_3^{-1} \\ -G_3 V_2 G_0 G_1^{-1} & G_3 V_2 G_0 G_2^{-1} & 1 \end{pmatrix}.$$

Obviously the homogeneous system associated with the set (1.137) has not only the desired physical solution $|\Psi^{(i)}\rangle = |\Psi_{bound}\rangle$ as a result of the second factor on the rhs of (3.140), but may have also solutions caused by M:

$$\underset{\sim}{M}\Theta = 0.$$

$$(3.141)$$

Those solutions related to Θ, if they exist, are non-physical and are called spurious. If we put

$$\begin{pmatrix} |\Theta_1\rangle \\ |\Theta_2\rangle \\ |\Theta_3\rangle \end{pmatrix} = \begin{pmatrix} G_1 V_3 |\chi_1\rangle \\ G_2 V_1 |\chi_2\rangle \\ G_3 V_2 |\chi_3\rangle \end{pmatrix}$$

$$(3.142)$$

(3.141) turns into the simple set

$$\begin{pmatrix} 1 & G_0 V_1 & -G_0 V_2 \\ -G_0 V_3 & 1 & G_0 V_2 \\ G_0 V_3 & -G_0 V_1 & 1 \end{pmatrix} \begin{pmatrix} |\chi_1\rangle \\ |\chi_2\rangle \\ |\chi_3\rangle \end{pmatrix} = 0.$$

$$(3.143)$$

Nontrivial solutions to that set have been proved to exist in a special case by *Federbush* [3.27].

We note that the kernel of (3.143) is connected after one iteration. Therefore its spectrum is discrete (see the discussion in Sect. 3.4.3) and the eigenvalues 1, required by (3.143), can occur only at discrete energies. They will in general be complex. If they occur in the neighbourhood of real energies which are of physical interest, their existence may require special numerical care. Regarding the example (3.143) where pair interactions occur with different signs, one can expect that the spurious energies are more sensitive to variations in the approximation scheme and potential parameters than the physical quantities, which may help to identify them.

Nevertheless, this example demonstrates that hunting just for connected kernels can lead to formulations which are no longer strictly equivalent to the underlying Schrödinger equation. They allow for additional, nonphysical, spurious solutions of the associated homogeneous problem.

We conclude this section with another example: the Weinberg equation [3.28] for the scattering state. Let us regard the resolvent identity

$$G = G_0 + G_0 \sum V_\beta G . \tag{3.144}$$

We want to separate in a systematic manner, the disconnected parts on the right hand side. The subseries in G containing only V_β can be collected through

$$G = G_\beta + G_\beta V^\beta G \tag{3.145}$$

and we get

$$G = G_0 + G_0 \sum V_\beta G_\beta + G_0 \sum V_\beta G_\beta V^\beta G . \tag{3.146}$$

The three terms are obviously of increasing connectivity ending with a completely connected one. We use this decomposition of G in (3.24) and get an integral equation for the scattering state [3.28]:

$$|\Psi_\alpha^{(+)}\rangle = |\phi_\alpha\rangle + G_0 \sum_\beta V_\beta G_\beta V^\beta |\Psi_\alpha^{(+)}\rangle . \tag{3.147}$$

It has a connected kernel but is not in unique correspondence to the underlying Schrödinger equation. Indeed, one has the factorization property [3.29]

$$(1 - G_0 \sum V_\beta G_\beta V^\beta) = (1 + \sum G_\beta V_\beta)(1 - G_0 \sum V_\gamma) , \tag{3.148}$$

which demonstrates explicitly that the homogeneous equation associated with (3.147) has not only bound state solutions corresponding to the Schrödinger equation

$$(1 - G_0 \sum V_\gamma)|\Psi\rangle = 0 \tag{3.149}$$

but also spurious ones occurring through

$$\left(1 + \sum_\beta G_p V_\beta\right)|\Theta\rangle = 0 . \tag{3.150}$$

Equipped with the Faddeev technique, we can apply it to the spurious problem (3.150) as well. Thus we introduce components $|\theta_\beta\rangle$ through

$$|\Theta\rangle = - \sum G_\beta V_\beta |\Theta\rangle \equiv - \sum |\theta_\beta\rangle \tag{3.151}$$

or

$$|\theta_\beta\rangle = -G_0 t_\beta \sum_\alpha |\theta_\alpha\rangle . \tag{3.152}$$

Then we remove the disconnected parts through

$$(1 + G_0 t_\beta)|\theta_\beta\rangle = -G_0 t_\beta \sum_{\alpha \neq \beta} |\theta_\alpha\rangle \tag{3.153}$$

and inversion to get

$$|\theta_\beta\rangle = -G_0 V_\beta \sum_{\alpha \neq \beta} |\theta_\alpha\rangle . \tag{3.154}$$

These are "spurious" Faddeev equations which are to be contrasted with the physical homogeneous ones

$$|\psi_\beta\rangle = G_0 t_\beta \sum_{\alpha \neq \beta} |\psi_\alpha\rangle . \tag{3.155}$$

Clearly this spurious problem is closely related to (3.143) which showed up within the channel array coupling scheme.

Equivalent to the cluster decomposition (3.146) is the insertion of Lippmann-Schwinger equations into each other. In addition it reveals at the same time the spurious multiplier. The state $|\Psi_\alpha(\varepsilon)\rangle \equiv i\varepsilon G(E+i\varepsilon)|\phi_\alpha\rangle$, $\varepsilon \neq 0$, obeys the Lippmann-Schwinger equations

$$|\Psi_\alpha(\varepsilon)\rangle - G_0 \sum_\mu V_\mu |\Psi_\alpha(\varepsilon)\rangle = i\varepsilon G_0 |\phi_\alpha\rangle \tag{3.156}$$

$$|\Psi_\alpha(\varepsilon)\rangle - G_\mu V^\mu |\Psi_\alpha(\varepsilon)\rangle = i\varepsilon G_\mu |\phi_\alpha\rangle . \tag{3.157}$$

Inserting (3.157) into (3.156) yields

$$|\Psi_\alpha(\varepsilon)\rangle - G_0 \sum_\mu V_\mu G_\mu V^\mu |\Psi_\alpha(\varepsilon)\rangle = (1 + \sum G_\mu V_\mu) i\varepsilon G_0 |\phi_\alpha\rangle . \tag{3.158}$$

In the limit $\varepsilon \to 0$ the rhs reduces to $|\phi_\alpha\rangle$ and we get the Weinberg equation (3.147). Comparing the rhs of (3.156) and (3.158) we can read off directly the factorization property (3.148). Though the insertion of valid equations into each other looks harmless at first sight, it introduces additional spurious solutions.

Further aspects and many more example of spurious solutions can be found in [3.26, 30].

3.4 Transition Operators

3.4.1 AGS-Equations

In a scattering problem, not the wavefunction, but the transition amplitudes (3.31)

$$\langle \phi_\beta | U_{\beta\alpha} | \phi_\alpha \rangle \equiv \langle \phi_\beta | V^\beta | \Psi_\alpha^{(+)} \rangle \tag{3.159}$$

are of central interest. The introduction of transition operators $U_{\beta\alpha}$ obviously parallels what is done in potential scattering:

$$\langle p' | t | p \rangle \equiv \langle p' | V | p \rangle^{(+)}. \tag{3.160}$$

The amplitudes in (3.159) cover all possible types of processes: $2 \to 2$ ($\alpha, \beta = 1, 2, 3$), $2 \to 3$ ($\alpha = 1, 2, 3$; $\beta = 0$), $3 \to 2$ ($\alpha = 0$, $\beta = 1, 2, 3$), and $3 \to 3$ ($\alpha = \beta = 0$). The corresponding channel states are defined in (3.8, 9).

As in the two-body problem one would like to work directly with integral equations for the transition operators. They are established very easily using the fundamental set of Lippmann-Schwinger equations which we reproduce for the sake of clarity. Specifically choose channel 1 as the initial channel and explicitly include the three-body force for our later convenience. Then $|\Psi_1^{(+)}\rangle$ is defined by

$$|\Psi_1^{(+)}\rangle = \phi_1 + G_1(V_2 + V_3 + V_4)|\Psi_1^{(+)}\rangle$$
$$|\Psi_1^{(+)}\rangle = \quad\quad G_2(V_3 + V_1 + V_4)|\Psi_1^{(+)}\rangle \tag{3.161}$$
$$|\Psi_1^{(+)}\rangle = \quad\quad G_3(V_1 + V_2 + V_4)|\Psi_1^{(+)}\rangle.$$

On the rhs, we recognize immediately the transition operators $U_{\beta 1}$, $\beta = 1, 2, 3$, we are looking for. Let us first ignore the three-body force V_4. Then we can read off, very naturally, three coupled equations for $U_{\beta 1}$:

$$U_{11}|\phi_1\rangle \equiv (V_2 + V_3)|\Psi_1^{(+)}\rangle = V_2 G_2 U_{21}|\phi_1\rangle + V_3 G_3 U_{31}|\phi_1\rangle$$
$$U_{21}|\phi_1\rangle \equiv (V_3 + V_1)|\Psi_1^{(+)}\rangle = V_3 G_3 U_{31}|\phi_1\rangle + V_1|\phi_1\rangle + V_1 G_1 U_{11}|\phi_1\rangle$$
$$U_{31}|\phi_1\rangle \equiv (V_1 + V_2)|\Psi_1^{(+)}\rangle = V_1|\phi_1\rangle + V_1 G_1 U_{11}|\phi_1\rangle + V_2 G_2 U_{21}|\phi_1\rangle. \tag{3.162}$$

It is usual to drop $|\phi_1\rangle$, and to keep in mind that the operators should be applied on the corresponding channel state. Moreover, one may put

$$V_1|\phi_1\rangle = G_0^{-1}|\phi_1\rangle, \tag{3.163}$$

which will be convenient in the context of the multiple scattering series and the Lovelace equations (see Sect. 3.4.3 and Sect. 3.6). Then the coupled set (3.162) achieves its standard form, known as the AGS-equations [3.31]:

$$U_{\beta\alpha} = \bar{\delta}_{\alpha\beta} G_0^{-1} + \sum_{\gamma} \bar{\delta}_{\gamma\beta} V_{\gamma} G_{\gamma} U_{\gamma\alpha} . \tag{3.164}$$

We use the very convenient notation

$$\bar{\delta}_{\alpha\beta} \equiv 1 - \delta_{\alpha\beta} . \tag{3.165}$$

The AGS-equations follow quite naturally from the fundamental set of Lippmann-Schwinger equations. They connect the operators for elastic, $U_{\alpha\alpha}$, and rearrangement processes, $U_{\gamma\alpha}$, $\gamma \neq \alpha$, $(\alpha, \gamma = 1, 2, 3)$. The operation $V_{\gamma} G_{\gamma}$ however requires $U_{\gamma\alpha}$ in the full Hilbert space and not only the matrix elements in the subspace of channel states for two-body fragmentations. As a consequence, the set (3.164) incorporates the break-up processes, too. The break-up amplitude (a process $2 \to 3$) is given as

$$\langle \phi_0 | U_{0\alpha} | \phi_\alpha \rangle \equiv \langle \phi_0 | \sum_{\gamma} V_{\gamma} | \Psi_\alpha^{(+)} \rangle . \tag{3.166}$$

We use again the fundamental set (3.161) and express the rhs as

$$\langle \phi_0 | U_{0\alpha} | \phi_\alpha \rangle = \langle \phi_0 | \left(\sum_{\gamma} V_{\gamma} (|\phi_\gamma\rangle \delta_{\gamma\alpha} + G_{\gamma} U_{\gamma\alpha} | \phi_\alpha \rangle \right) . \tag{3.167}$$

The first term vanishes on-shell and we define the break-up operator as

$$U_{0\alpha} = \sum_{\gamma} V_{\gamma} G_{\gamma} U_{\gamma\alpha} . \tag{3.168}$$

Thus, indeed, the information stored in the three operators $U_{\gamma\alpha}$, $\gamma = 1, 2, 3$ is sufficient to calculate the break-up process.

We can look closer and find yet another form for the break-up operator, which occurs directly within (3.164). Unfolding $\bar{\delta}_{\gamma\beta}$ (3.164) can be rewritten as

$$(1 + V_\beta G_\beta) U_{\beta\alpha} = \bar{\delta}_{\alpha\beta} G_0^{-1} + \sum_{\gamma} V_{\gamma} G_{\gamma} U_{\gamma\alpha} . \tag{3.169}$$

Again considering only the on-shell break-up amplitude, we find the form

$$U_{0\alpha} = (1 + V_\beta G_\beta) U_{\beta\alpha} . \tag{3.170}$$

Consequently, the amplitude can also be written as

$$\langle \phi_0|U_{0\alpha}|\phi_\alpha\rangle = \langle \phi_0|(1 + V_\beta G_\beta)U_{\beta\alpha}|\phi_\alpha\rangle$$
$$= \langle \phi_\beta^{(-)}|U_{\beta\alpha}|\phi_\alpha\rangle . \tag{3.171}$$

In the last form the state $|\phi_\beta^{(+)}\rangle = |p_\beta\rangle^{(+)}|q_\beta\rangle$ of (3.12) and (3.54) describes the motion of three unbound particles where the pair interaction V_β, however, is included. Let us call them distorted channel states. Since the spectral representation of G_γ includes both channel states $|\phi_\gamma\rangle$ and these types of scattering states $|\phi_\gamma^{(\pm)}\rangle$, we see that in the midst of (3.164) the break-up operators are already present.

For later use we derive equations for operators describing the processes of the type $3 \rightarrow 2$. According to (3.159) the transition amplitudes are

$$\langle \phi_\alpha|U_{\alpha 0}|\phi_0\rangle \equiv \langle \phi_\alpha|V^\alpha|\Psi_0^{(+)}\rangle . \tag{3.172}$$

Here we use the fundamental set of Lippmann-Schwinger equations (3.55) for $|\Psi_0^{(+)}\rangle$:

$$|\Psi_0^{(+)}\rangle = |\phi_\gamma^{(+)}\rangle + G_\gamma V^\gamma|\Psi_0^{(+)}\rangle, \quad \gamma = 1, 2, 3 \tag{3.173}$$

and derive in the same manner as for (3.162)

$$U_{\alpha 0} = \sum_\beta \bar{\delta}_{\beta\alpha} t_\beta + \sum_\beta \bar{\delta}_{\beta\alpha} V_\beta G_\beta U_{\beta 0} . \tag{3.174}$$

Note that, in contrast to (3.164), each equation has its driving term, which arises through

$$V_\gamma|\phi_\gamma^{(+)}\rangle = t_\gamma|\phi_0\rangle . \tag{3.175}$$

Let us conclude this first part and introduce the two-body transition operators. Equivalent to (3.96) one has

$$V_\gamma G_\gamma = t_\gamma G_0 \tag{3.176}$$

and (3.164) reads

$$\begin{pmatrix} U_{1\alpha} \\ U_{2\alpha} \\ U_{3\alpha} \end{pmatrix} = \begin{pmatrix} \bar{\delta}_{\gamma\alpha} G_0^{-1} \end{pmatrix} + \begin{pmatrix} 0 & t_2 & t_3 \\ t_1 & 0 & t_3 \\ t_1 & t_2 & 0 \end{pmatrix} G_0 \begin{pmatrix} U_{1\alpha} \\ U_{2\alpha} \\ U_{3\alpha} \end{pmatrix} . \tag{3.177}$$

Comparing the kernels in (3.100) and (3.177) we note that they are just transposed to each other. This will be the decisive key to understanding the convergence or divergence properties of the multiple scattering series in Sect.

3.4.3. Clearly the kernel in (3.177) also has the property of becoming connected after one iteration.

Let us now reintroduce the three-body force.

One can think of many possible ways to incorporate V_4. We may write for instance

$$V_2 + V_3 + V_4 = (V_2 + \tfrac{1}{2}V_4) + (V_3 + \tfrac{1}{2}V_4) \tag{3.178}$$

and proceed as above. In this manner the first equation in (3.162) would be replaced by

$$(V_2 + V_3 + V_4)|\Psi_1^{(+)}\rangle \equiv U_{11}|\phi_1\rangle = (V_2 + \tfrac{1}{2}V_4)G_2 U_{21}|\phi_1\rangle$$
$$+ (V_3 + \tfrac{1}{2}V_4)G_3 U_{31}|\phi_1\rangle . \tag{3.179}$$

Another possibility which we shall use for the bound state is to split V_4 into three terms:

$$V_4 = \sum_{i=1}^{3} V_4^{(i)} . \tag{3.180}$$

For certain three-body force models this is very natural (see Sect. 3.4.4). Then one would group

$$V_2 + V_3 + V_4 = (V_2 + V_4^{(2)}) + (V^3 + V_4^{(3)}) + V_4^{(1)} \tag{3.181}$$

and (3.162) would be replaced by

$$(V_2 + V_3 + V_4)|\Psi_1^{(+)}\rangle \equiv U_{11}|\phi_1\rangle = (V_2 + V_4^{(2)})G_2 U_{21}|\phi_1\rangle$$
$$+ (V_3 + V_4^{(3)})G_3 U_{31}|\phi_1\rangle + V_4^{(1)}|\phi_1\rangle + V_4^{(1)}G_1 U_{11}|\phi_1\rangle . \tag{3.182}$$

In each case we end up again with 3 coupled equations for $U_{\beta\alpha}$.

We shall consider now in some detail another possibility [3.32], in which we introduce a fourth transition operator

$$U_{41}|\phi_1\rangle \equiv (V_1 + V_2 + V_3)|\Psi_1^{(+)}\rangle \tag{3.183}$$

together with a fourth equation for $\Psi_1^{(+)}$:

$$|\Psi_1^{(+)}\rangle = G_4(V_1 + V_2 + V_3)|\Psi_1^{(+)}\rangle . \tag{3.184}$$

The Green's operator

$$G_4 = (E + i0 - H_0 - V_4)^{-1} \tag{3.185}$$

obeys the resolvent identity

$$G = G_4 + G_4(V_1 + V_2 + V_3)G.$$ (3.186)

Since

$$\lim_{\varepsilon \to 0} i\varepsilon G_4 \phi_1 = 0,$$ (3.187)

which should be obvious remembering the discussion in Sect. 3.2, the use of (3.186) in (3.24) yields immediately (3.184). With the extension (3.183) and (3.184), we can easily generalise the derivation of the set (3.164) in the following manner. We build up $U_{11}|\phi_1\rangle$ as

$$(V_2 + V_3 + V_4)|\Psi_1^{(+)}\rangle \equiv U_{11}|\phi_1\rangle = V_2 G_2 U_{21}|\phi_1\rangle$$
$$+ V_3 G_3 U_{31}|\phi_1\rangle + V_4 G_4 U_{41}|\phi_1\rangle$$ (3.188)

and correspondingly the expressions for $U_{21}|\phi_1\rangle$ and $U_{31}|\phi_1\rangle$. The fourth operator $U_{41}|\phi_1\rangle$ can be expressed through U_{11}, U_{21}, and U_{31} as

$$U_{41}|\phi_1\rangle = V_1|\phi_1\rangle + V_1 G_1 U_{11}|\phi_1\rangle + V_2 G_2 U_{21}|\phi_1\rangle + V_3 G_3 U_{31}|\phi_1\rangle.$$ (3.189)

Altogether we get

$$U_{\beta\alpha} = \bar{\delta}_{\beta\alpha} G_0^{-1} + \sum_\gamma \bar{\delta}_{\gamma\beta} V_\gamma G_\gamma U_{\gamma\alpha} + V_4 G_4 U_{4\alpha}$$
$$U_{4\alpha} = G_0^{-1} + \sum_\gamma V_\gamma G_\gamma U_{\gamma\alpha}.$$ (3.190)

The additional operators $U_{4\alpha}$ are not only auxiliary quantities, but have physical meaning. They all produce physical transition amplitudes, as we shall explain now. We have already encountered one case, (3.171), where a break-up operator could be presented in two forms, depending on whether undistorted or distorted channel states are used. In the same spirit we can introduce into the description of a break-up configuration scattering states

$$|\phi_4^{(\pm)}\rangle = \lim_{\varepsilon \to 0} i\varepsilon G_4(E \pm i\varepsilon)|\phi_0\rangle.$$ (3.191)

They obey the equations

$$|\phi_4^{(\pm)}\rangle = |\phi_0\rangle + G_4^{(\pm)} V_4|\phi_0\rangle$$ (3.192)

$$|\phi_4^{(\pm)}\rangle = |\phi_0\rangle + G_0^{(\pm)} V_4|\phi_4^{(\pm)}\rangle.$$ (3.193)

Equipped with these distorted break-up states we can rewrite the break-up amplitude:

$$\langle \phi_0 | V^0 | \Psi_\alpha^{(\pm)} \rangle \equiv \langle \phi_0 | U_{0\alpha} | \phi_\alpha \rangle$$
$$= \langle \phi_4^{(-)} | U_{4\alpha} | \phi_\alpha \rangle = \langle \phi_0 | (1 + V_4 G_4) U_{4\alpha} | \phi_\alpha \rangle . \tag{3.194}$$

Indeed, according to (3.190),

$$(1 + V_4 G_4) U_{4\alpha} | \phi_\alpha \rangle = (G_0^{-1} + \sum_\gamma V_\gamma G_\gamma U_{\gamma\alpha} + V_4 G_4 U_{4\alpha}) | \phi_\alpha \rangle$$
$$= (\sum V_\gamma + V_4) | \Psi_\alpha^{(+)} \rangle \equiv U_{0\alpha} | \phi_\alpha \rangle . \tag{3.195}$$

One arrives at the same conclusion by regarding the homogeneous Lippmann-Schwinger equation (3.184) for the scattering state $|\Psi_1^{(+)}\rangle$. The spectral decomposition of G_4 in terms of the eigenstates $\phi_4^{(\pm)}$ of H_4 can be used to evaluate the asymptotic form of $|\Psi_1^{(+)}\rangle$ in the break-up channel. The resulting amplitude of the outgoing wave is $\langle \phi_4^{(-)} | U_{4\alpha} | \phi_\alpha \rangle$. We leave that study for the reader. There is another interesting point, however, connected to the homogeneous equation (3.184). The three-body force V_4 may alone support bound states, $|\phi_4\rangle$, at energies E_4. They show up as poles in the resolvent operator G_4 and yield a term

$$|\phi_4\rangle \frac{1}{E - E_4} \langle \phi_4 | V_1 + V_2 + V_3 | \Psi_1^{(+)} \rangle \tag{3.196}$$

on the rhs of (3.184). Clearly the scattering state $|\Psi_1^{(+)}(E)\rangle$ does not have a pole at $E = E_4$ and therefore

$$\langle \phi_4 | V_1 + V_2 + V_3 | \Psi_1^{(+)} \rangle |_{E=E_4} = 0 . \tag{3.197}$$

In other words the transition operators $U_{4\alpha}$ have the property

$$\langle \phi_4 | U_{4\alpha} | \phi_\alpha \rangle = 0 \quad \text{at} \quad E = E_4 . \tag{3.198}$$

Let us now regard the transitions initiated by 3 particles. The amplitudes are

$$\langle \phi_\alpha | V^\alpha | \Psi_0^{(+)} \rangle \equiv \langle \phi_\alpha | U_{\alpha 0} | \phi_0 \rangle , \quad \alpha = 0, 1, 2, 3 . \tag{3.199}$$

Introducing V_4 distortions to the initial state we have

$$\langle \phi_\alpha | U_{\alpha 0} | \phi_0 \rangle \equiv \langle \phi_\alpha | U_{\alpha 4} | \phi_4^{(+)} \rangle , \quad \alpha = 1, 2, 3 \tag{3.200}$$

and for $\alpha = 0$, with V_4 distortions also in the final state,

$$\langle\phi_0|U_{00}|\phi_0\rangle \equiv \langle\phi_4^{(-)}|U_{44}|\phi_4^{(+)}\rangle .\tag{3.201}$$

Instead of a straightforward argument, let us amuse ourselves by guessing the form which the equations for these operators will take. Regarding (3.190), it is tempting to supplement that set by a fourth column of operators defined by

$$U_{\alpha 4} = \bar{\delta}_{\alpha 4}G_0^{-1} + \sum_\gamma \bar{\delta}_{\gamma\alpha}V_\gamma G_\gamma U_{\gamma 4} + V_4 G_4 U_{44}$$

$$U_{44} = \sum_\gamma V_\gamma G_\gamma U_{\gamma 4} .\tag{3.202}$$

It is now time to introduce a better notation. We introduce indices a, b, c, which run over $\beta = 1, 2, 3$ and the value 4. Then a more compact notation for (3.202) is

$$U_{a4} = \bar{\delta}_{a4}G_0^{-1} + \sum_c \bar{\delta}_{ca}t_c G_0 U_{c4} ,\tag{3.203}$$

where

$$t_c G_0 = V_c G_c , \quad c = 1, 2, 3, 4 .\tag{3.203 a}$$

Let us iterate (3.202). One finds

$$U_{a4} = \bar{\delta}_{a4}G_0^{-1} + \sum_c \bar{\delta}_{ca}t_c\bar{\delta}_{c4} + \sum_{c_1 c_1} \bar{\delta}_{ca}t_c G_0 \bar{\delta}_{c_1 c}t_{c_1}\bar{\delta}_{c_1 4} + \cdots .\tag{3.204}$$

Thus $U_{\alpha 4}$ contains all operators which start with $t_c \neq t_4$ on the right and end with $t_c \neq t_\alpha$. Therefore we guess that the matrix element $\langle\phi_\alpha | U_{\alpha 4} | \phi_4^{(+)}\rangle$ describes the processes initiated by 3 particles and ending in two fragment channels. Before consolidating that claim we point to another interesting mechanism. The operator $(1 + G_0 t_4)$ adds the missing t_4 to the right in each term of the perturbation series

$$U_{a4}(1 + G_0 t_4) = \bar{\delta}_{a4}G_0^{-1} + \sum_c \bar{\delta}_{ac}t_c + \sum_{cc_1}\bar{\delta}_{ac}t_c G_0 t_{c_1} + \cdots .\tag{3.205}$$

Following the same line of reasoning this product of operators should be applied to a free state and the matrix element $(\phi_\alpha | U_{\alpha 4}(1 + G_0 t_4) | \phi_0)$ should represent the same physical processes as the matrix element $(\phi_\alpha | U_{\alpha 4} | \phi_4^{(+)})$. That the two matrix elements are equal follows directly from (3.192).

Let us now finally establish the link to our general notation. The processes initiated by three particles are described by the amplitudes

$$\langle\phi_\alpha|V^\alpha|\Psi_0^{(+)}\rangle \equiv \langle\phi_\alpha|U_{\alpha 0}|\phi_0\rangle, \quad \alpha = 0, 1, 2, 3 .\tag{3.206}$$

Again we supplement the set of three Lippmann-Schwinger equations (3.173) for the state $|\Psi_0^{(+)}\rangle$ by a fourth equation, which is now inhomogenous

$$|\Psi_0^{(+)}\rangle = |\phi_4^{(+)}\rangle + G_4 V^4 |\Psi_0^{(+)}\rangle . \tag{3.207}$$

Thus we can quite simply generalize (3.174) as

$$U_{a0} = \sum_c \bar{\delta}_{ac} t_c + \sum_c \bar{\delta}_{ac} t_c G_0 U_{c0} . \tag{3.208}$$

In order to establish (3.200) and (3.201), we have to compare the coupled sets (3.203) for U_{a4} and (3.208) for U_{a0}. Comparison of the matrix element $(\phi_\alpha | U_{a0} | \phi_0)$ with $(\phi_\alpha | U_{a4}(1 + G_0 t_4) | \phi_0)$ suggests the connection

$$U_{a0} = U_{a4}(1 + G_0 t_4) - \bar{\delta}_{a4} G_0^{-1} . \tag{3.209}$$

This can be easily proven by direct substitution of (3.203) into (3.209). This verifies the on-shell relations (3.200) and (3.201).

In compact notation (3.190) and (3.203) read

$$U_{ab} = \bar{\delta}_{ab} G_0^{-1} + \sum_c \bar{\delta}_{ac} V_c G_c U_{cb} . \tag{3.210}$$

A set of equations of this type has been proposed in [3.33], however with different driving terms, corresponding to two different choices of off-shell transition operators. Related to that form is the one given in [3.34], where the V_4 interaction is not summed up beforehand and therefore $t_4 \equiv V_4$ occurs in the kernel.

In a usual experiment the initial channel contains only two fragments, and one is interested only in the $U_{\alpha\beta}$. Also the operator $U_{4\beta}$, $\beta = 1, 2, 3$ may be trivially eliminated and (3.210) appears again as a set of three equations

$$U_{\alpha\beta} = (\bar{\delta}_{\alpha\beta} + t_4 G_0) G_0^{-1} + \sum_\gamma (\bar{\delta}_{\gamma\alpha} + t_4 G_0) V_\gamma G_\gamma U_{\gamma\beta} . \tag{3.211}$$

The break-up operator can then be calculated according to (3.195) and (3.210) as

$$U_{0\beta} = t_4 + (1 + t_4 G_0) \sum_\gamma t_\gamma G_0 U_{\gamma\beta} \tag{3.212}$$

or

$$U_{0\beta} = (1 + t_a G_0) U_{\alpha\beta} . \tag{3.213}$$

3.4.2 Unitarity

The scattering process initiated in a channel α will lead in general to various final channels β. We expect that the probabilities for scattering into all open final channels will sum up to 1. We shall prove this in two ways.

As in Sect. 1.5, we shall introduce another type of solutions to the time dependent Schrödinger equation, states $|\Psi_\alpha^{(-)}\rangle$. They are defined to coincide with channel states in the infinite future:

$$\lim_{t\to\infty} \| \Psi_\alpha^{(-)}(t) - \phi_\alpha(t)\| = 0 \tag{3.214}$$

or

$$|\Psi_\alpha^{(-)}(0)\rangle = \lim_{t\to+\infty} e^{iHt}e^{-iH_\alpha t}|\phi_\alpha(0)\rangle \equiv \Omega_\alpha^{(-)}|\phi_\alpha(0)\rangle . \tag{3.215}$$

Then we can express the probability amplitude for the transition from channel α to channel β as in potential scattering by

$$\begin{aligned}
A_{\beta\alpha} &= \lim_{t\to+\infty} \langle\phi_\beta(t)|\Psi_\alpha^{(+)}(t)\rangle \\
&= \lim_{t\to+\infty} \langle\Psi_\beta^{(-)}(t)|\Psi_\alpha^{(+)}(t)\rangle \\
&= \langle\Psi_\beta^{(-)}(0)|\Psi_\alpha^{(+)}(0)\rangle .
\end{aligned} \tag{3.216}$$

Since the channel Möller wave operators $\Omega_\alpha^{(\pm)}$ map the channel states ϕ_α into $|\Psi_\alpha^{(\pm)}\rangle$ we can interpret $A_{\beta\alpha}$ as matrix elements of the set of S-operators

$$\hat{S}_{\beta\alpha} \equiv \Omega_\beta^{(-)+}\,\Omega_\alpha^{(+)} \tag{3.217}$$

or

$$A_{\beta\alpha} = \langle\phi_\beta|\hat{S}_{\beta\alpha}|\phi_\alpha\rangle \equiv S_{\beta\alpha}. \tag{3.218}$$

The set of states $\{|\Psi_\alpha^{(+)}\rangle,\ \alpha = 0, 1, 2, 3\}$ and $\{|\Psi_\alpha^{(-)}\rangle,\ \alpha = 0, 1, 2, 3\}$ span the same space, namely the part of the total three body Hilbert space, which is orthogonal to the three-body bound state. Therefore we can expand $|\Psi_\alpha^{(+)}\rangle$ in terms of all the states $|\Psi_\alpha^{(-)}\rangle$ and vice versa:

$$\begin{aligned}
|\Psi_\alpha^{(+)}\rangle &= \sum_\gamma \!\!\!\!\!\!\int |\Psi_\gamma^{(-)}\rangle\langle\Psi_\gamma^{(-)}|\Psi_\alpha^{(+)}\rangle \\
\\
&= \sum_\gamma \!\!\!\!\!\!\int |\Psi_\gamma^{(-)}\rangle S_{\gamma\alpha} .
\end{aligned} \tag{3.219}$$

Since the states $|\Psi_\alpha^{(+)}\rangle$ and $|\Psi_\alpha^{(-)}\rangle$ are orthogonal among themselves [see (3.29, 30) in the case of $|\Psi_\alpha^{(+)}\rangle$] it follows that

$$\begin{aligned}
\langle\Psi_{\alpha'}^{(+)}|\Psi_\alpha^{(+)}\rangle = \delta(\alpha'-\alpha) &= \sum_\gamma\sum_{\gamma'} \!\!\!\!\!\!\int \langle\Psi_{\gamma'}^{(-)}|\Psi_\gamma^{(-)}\rangle S_{\gamma'\alpha}^* S_{\gamma\alpha} \\
&= \sum_\gamma \!\!\!\!\!\!\int S_{\gamma\alpha'}^* S_{\gamma\alpha} .
\end{aligned} \tag{3.220}$$

Expansion in the reversed order yields

$$\sum_{\gamma} S^*_{\beta'\gamma} S_{\beta\gamma} = \delta(\beta - \beta') \, . \tag{3.221}$$

These are the unitarity relations for the S-matrix elements. Specifically choosing $\alpha = \alpha'$ in (3.220), we see that the transition probabilities from channel α into all channels γ indeed sum up to 1.

How are the S-matrix elements $S_{\beta\alpha}$ related to the transition amplitudes $\langle \phi_\beta | U_{\beta\alpha} | \phi_\alpha \rangle$? We shall recover the same structural link as in potential scattering. Let us first regard transitions between two-body fragmentation channels and let us replace the schematic notation by an explicit one:

$$S_{q'_\alpha q_\beta} \equiv \langle \Psi^{(-)}_{q'_\alpha} | \Psi^{(+)}_{q_\beta} \rangle \, . \tag{3.222}$$

It is now convenient to use the relation between $|\Psi^{(+)}_{q_\alpha}\rangle$ and $|\Psi^{(-)}_{q_\alpha}\rangle$. The two stationary states obey

$$|\Psi^{(\pm)}_{q_\alpha}\rangle = |\phi_{q_\alpha}\rangle + (E_{q_\alpha} \pm i0 - H)^{-1} V^\alpha |\phi_{q_\alpha}\rangle \, . \tag{3.223}$$

Therefore we deduce that

$$|\Psi^{(-)}_{q_\alpha}\rangle - |\Psi^{(+)}_{q_\alpha}\rangle = 2 i \pi \, \delta(E_{q_\alpha} - H) V^\alpha |\phi_{q_\alpha}\rangle \tag{3.224}$$

and we can rewrite $S_{q'_\alpha q_\beta}$ as

$$\begin{aligned}
S_{q'_\alpha q_\beta} &= \langle \Psi^{(+)}_{q'_\alpha} | \Psi^{(+)}_{q_\beta} \rangle - 2 i \pi \, \delta(E_{q'_\alpha} - E_{q_\beta}) \langle \phi_{q_\alpha} | V^\alpha | \Psi^{(+)}_{q_\beta} \rangle \\
&= \delta_{\alpha\beta} \delta^3 (q'_\alpha - q_\beta) - 2 i \pi \, \delta(E_{q'_\alpha} - E_{q_\beta}) \langle \phi_{q'_\alpha} | U_{\alpha\beta} | \phi_{q_\beta} \rangle \, .
\end{aligned} \tag{3.225}$$

The first term results from the orthogonality relations (3.29, 30) of the states $|\Psi^{(+)}\rangle$ and corresponds to the free motion of the two initial fragments. The second term incorporates the transitions and exhibits energy conservation. Clearly the same steps go through for $\alpha = 0$, the break-up channel.

The unitarity relations for $S_{\beta\alpha}$ now impose certain restrictions on the physical on-the-energy-shell transition amplitudes $\langle \phi_\alpha | U_{\alpha\beta} | \phi_\beta \rangle$. We insert (3.225) into (3.220) and find the on-shell relations:

$$\begin{aligned}
\delta(E_{q'_\alpha} - E_{q_\alpha}) &\Big[i \langle \phi_{q'_\alpha} | U_{\alpha'\alpha} | \phi_{q_\alpha} \rangle - i \langle \phi_{q_\alpha} | U_{\alpha\alpha'} | \phi_{q'_\alpha} \rangle^* \\
&- 2\pi \sum_{\gamma=1,2,3} \int dq_\gamma \, \delta(E_{q_\gamma} - E_{q_\alpha}) \langle \phi_{q_\gamma} | U_{\gamma\alpha'} | \phi_{q'_\alpha} \rangle^* \langle \phi_{q_\gamma} | U_{\gamma\alpha} | \phi_{q_\alpha} \rangle \\
&- 2\pi \int dp \, dq \, \delta(E_{pq} - E_{q_\alpha}) \langle pq | U_{0\alpha'} | \phi_{q'_\alpha} \rangle^* \langle pq | U_{0\alpha} | \phi_{q_\alpha} \rangle \Big] = 0 \, .
\end{aligned} \tag{3.226}$$

It is also an instructive exercise to derive the unitarity relations directly from the AGS-equations, which after all define them. Let us write the spectral decomposition of the resolvent operator G_c in symbolic manner as

$$G_c = \int |c\rangle \frac{1}{e_c} \langle c|. \tag{3.227}$$

This exhibits the cut structure of the kernel in (3.210):

$$U_{\alpha\beta}|\phi_\beta\rangle = \bar{\delta}_{\alpha\beta} G_0^{-1}|\phi_\beta\rangle + \sum_c \int \bar{\delta}_{ac} \langle \phi_\alpha |V_c|c\rangle \frac{1}{e_c} \langle c|U_{c\beta}|\phi_\beta\rangle . \tag{3.228}$$

If we would choose an energy $E - i0$, instead of $E + i0$ as in (3.228), different amplitudes would be defined. We distinguish them by explicitly inserting the energy arguments. The amplitudes defined on the lower rims of the continuum cuts satisfy the equations

$$U_{\alpha\beta}(E - i0)|\phi_\beta\rangle = \bar{\delta}_{\alpha\beta} G_0^{-1}|\phi_\beta\rangle + \sum_c \int \bar{\delta}_{ac} V_c |c\rangle \frac{1}{e_c^{(-)}} \langle c|U_{c\beta}|E - i0)|\phi_\beta\rangle . \tag{3.229}$$

We can convince ourselves that $U_{\alpha\beta} \equiv U_{\alpha\beta}(E + i0)$ and $U_{\alpha\beta}(E - i0)$ are really different by subtracting (3.228) from (3.229):

$$U_{\alpha\beta}(E - i0)|\phi_\beta\rangle - U_{\alpha\beta}(E + i0)|\phi_\beta\rangle$$

$$= \sum_c \bar{\delta}_{ac} \int V_c|c\rangle \frac{1}{e_c^{(-)}} \langle c|U_{c\beta}(E - i0)|\phi_\beta\rangle$$

$$\quad - \sum_c \bar{\delta}_{ac} \int V_c|c\rangle \frac{1}{e_c} \langle c|U_{c\beta}(E + i0)|\phi_\beta\rangle$$

$$= \sum_c \int \bar{\delta}_{ac} V_c|c\rangle \frac{1}{e_c^{(-)}} (\langle c|U_{c\beta}(E - i0)|\phi_\beta\rangle - \langle c|U_{c\beta}(E + i0)|\phi_\beta\rangle)$$

$$\quad + \sum_c \int \bar{\delta}_{ac} V_c|c\rangle \left(\frac{1}{e_c^{(-)}} - \frac{1}{e_c} \right) \langle c|U_{c\beta}(E + i0)|\phi_\beta\rangle . \tag{3.230}$$

We get a nonvanishing contribution in the second term of the last equation for intermediate energies, E_c, which contribute to

$$\frac{1}{e_c^{(-)}} - \frac{1}{e_c} \equiv \frac{1}{E - i0 - E_c} - \frac{1}{E_c + i0 - E_c} = 2i\pi \, \delta(E - E_c) . \tag{3.231}$$

In order words the intermediate energies have to be on-shell. Equation (3.230) can be considered as integral equations for the difference

$$X_a \equiv U_{\alpha\beta}(E - i0)|\phi_\beta\rangle - U_{\alpha\beta}(E + i0)|\phi_\beta\rangle . \tag{3.232}$$

We put it in a form which suggests its solution:

$$X_a - \sum_c \bar{\delta}_{ac} V_c G_c^{(-)} X_c = \sum_{\gamma=1}^{3} \int \bar{\delta}_{a\gamma} V_\gamma |\phi_\gamma\rangle 2\pi i\, \delta(E - E_\gamma)\langle \phi_\gamma | U_{\gamma\beta}(E + i0) |\phi_\beta\rangle$$
$$+ \bar{\delta}_{a4} V_4 |\phi_4\rangle 2\pi i\, \delta(E - E_4)\langle \phi_4 | U_{4\beta}(E + i0)|\phi_\beta\rangle$$
$$+ \sum_c \int \bar{\delta}_{ac} V_c |\phi_c^{(-)}\rangle 2\pi i\, \delta(E - E_c)\langle \phi_c^{(-)}|U_{c\beta}(E + i0)|\phi_\beta\rangle. \tag{3.233}$$

First we note that the on-the-energy-shell transition into the three-body bound state $|\phi_4\rangle$ is zero as pointed out in (3.198). Secondly the integral equation (3.233) has the same kernel as in (3.229) or in (3.208) with $E + i0 \to E - i0$. Only the driving term consists of several contributions. The first term is a linear combination of driving terms present in (3.229). The third term is more tricky. We remember (3.171) and (3.213):

$$\langle \phi_c^{(-)}|U_{c\beta}(E + i0)|\phi_\beta\rangle = \langle \phi_0 |U_{0\beta}(E + i0)|\phi_\beta\rangle. \tag{3.233a}$$

Note that neither the free state $\langle \phi_0|$ nor its energy E_{pq} depend on the channel index c. Further, using (3.175), we can give the third term on the rhs of (3.233) the form

$$\sum_c \bar{\delta}_{ac} t_c^{(-)} \int |\phi_0\rangle 2\pi i\, \delta(E - E_{pq})\langle \phi_0 |U_{0\beta}(E + i0)|\phi_\beta\rangle. \tag{3.234}$$

This is obviously proportional to the driving term of the set (3.208) written for the operators $U_{\alpha 0}(E - i0)$.

Altogether we conclude that the solution of (3.233) must be a linear combination of the solutions of (3.208) and (3.229). Thus we end up with

$$\langle \phi_\alpha |U_{\alpha\beta}(E - i0)|\phi_\beta\rangle - \langle \phi_\alpha |U_{\alpha\beta}(E + i0)|\phi_\beta\rangle$$
$$= \sum_{\gamma=1}^{3} \int \langle \phi_\alpha |U_{\alpha\gamma}(E - i0)|\phi_\gamma\rangle 2\pi i\, \delta(E - E_\gamma)\langle \phi_\gamma |U_{\gamma\beta}(E + i0)|\phi_\beta\rangle$$
$$+ \int \langle \phi_\alpha |U_{\alpha 0}(E - i0)|\phi_0\rangle 2\pi i\, \delta(E - E_{pq})\langle \phi_0 |U_{0\beta}(E + i0)|\phi_\beta\rangle. \tag{3.235}$$

One would like to have an equation involving only the operators on the upper rim of the cuts. This is easily accomplished. All the matrix elements in (3.235) are on shell and are defined in (3.159) as

$$\langle \phi_\alpha |U_{\alpha\beta}(E \pm i0)|\phi_\beta\rangle \equiv \langle \phi_\alpha |V^\alpha |\Psi_\beta^{(\pm)}\rangle. \tag{3.236}$$

We insert the definition of $\Psi_\beta^{(\pm)}$, (3.223), and get

$$\langle \phi_\alpha |U_{\alpha\beta}(E - i0)|\phi_\beta\rangle = \langle \phi_\alpha |V^\alpha (1 + G^{(-)}V^\beta)|\phi_\beta\rangle$$
$$= \langle \phi_\alpha |V^\alpha |\phi_\beta\rangle + \langle \phi_\beta |V^\beta G^{(+)}V^\alpha |\phi_\alpha\rangle^*$$
$$= \langle \phi_\beta |V^\alpha - V^\beta |\phi_\alpha\rangle^* + \langle \phi_\beta |V^\beta |\Psi_\alpha^{(+)}\rangle^* \tag{3.237}$$
$$= \langle \phi_\beta |V_\beta - V_\alpha |\phi_\alpha\rangle^* + \langle \phi_\beta |U_{\beta\alpha}(E + i0)|\phi_\alpha\rangle^*.$$

The first term is zero up to oscillating surface terms in the case $\beta = 0$, which can be dropped. Therefore we find

$$\langle \phi_\beta | U_{\beta\alpha}(E + i0) | \phi_\alpha \rangle^* - \langle \phi_\alpha | U_{\alpha\beta}(E + i0) | \phi_\beta \rangle$$

$$= \sum_{\gamma=1}^{3} \int \langle \phi_\gamma | U_{\gamma\alpha}(E + i0) | \phi_\alpha \rangle^* \, 2\pi i \, \delta(E - E_\gamma) \langle \phi_\gamma | U_{\gamma\beta}(E + i0) | \phi_\beta \rangle$$

$$+ \int \langle \phi_0 | U_{0\alpha}(E + i0) | \phi_\alpha \rangle^* \, 2\pi i \, \delta(E - E_{pq}) \langle \phi_0 | U_{0\beta}(E + i0) | \phi_\beta \rangle , \quad (3.238)$$

which coincides with (3.226).

Clearly for $\alpha = \beta$ and $q_\alpha = q_\beta$, the lhs yields the imaginary part of the forward scattering amplitude, whereas the rhs sums up all transition probabilities initiated from channel α which is proportional to the total cross section. This is the content of the optical theorem.

3.4.3 Multiple Scattering Series

Scattering a neutron from a deuteron at high energies, we expect that the neutron hits one constituent of the deuteron just once and leaves. For lower energies the probability for two, three, or more collisions increases. In general we expect a sequence of multiple scattering events. Indeed the transition operator decomposes quite naturally into such a sequence of multiple collision processes as is seen by iterating the AGS-equations (3.210):

$$U_{\beta\alpha} = \bar{\delta}_{\beta\alpha} G_0^{-1} + \sum_c \bar{\delta}_{c\beta} t_c \bar{\delta}_{c\alpha} + \sum_c \bar{\delta}_{c\beta} t_c G_0 \sum_{c_1} \bar{\delta}_{c_1 c} t_{c_1} \bar{\delta}_{c_1 \alpha}$$

$$+ \sum_c \bar{\delta}_{c\beta} t_c G_0 \sum_{c_1} \bar{\delta}_{c_1 c} t_{c_1} G_0 \sum_{c_2} \bar{\delta}_{c_2 c_1} t_{c_2} \bar{\delta}_{c_2 \alpha} + \cdots . \quad (3.239)$$

The structure of the multiple scattering series is very transparent. Let us first regard transitions between two fragment channels β and α. Clearly the first two-body interaction to the right in each term has to take place between a pair of particles which is different from the "pair α". This restriction does not apply of course to the three-body interaction, which can act right away. Since the t_c's sum up all consecutive interactions V_c to all orders, two t-matrices with equal indices cannot follow each other. Finally the last interaction to the left cannot take place between the "pair β". These simple rules are obviously obeyed by the first few terms shown in (3.239). The first term in (3.239) is special and shows up only in the two rearrangement processes. We present graphically a few low order processes in Fig. 3.1, which may help to visualize the sequence of interactions and free intermediate propagations.

The first process in the series for the break-up amplitude, the first order term in t, is called the impulse approximation. Clearly in that approximation, one particle is unaffected and its momentum distribution in the final state is the same as in the bound pair-wave function of the initial state. For applications in the case of the deuteron see [3.35].

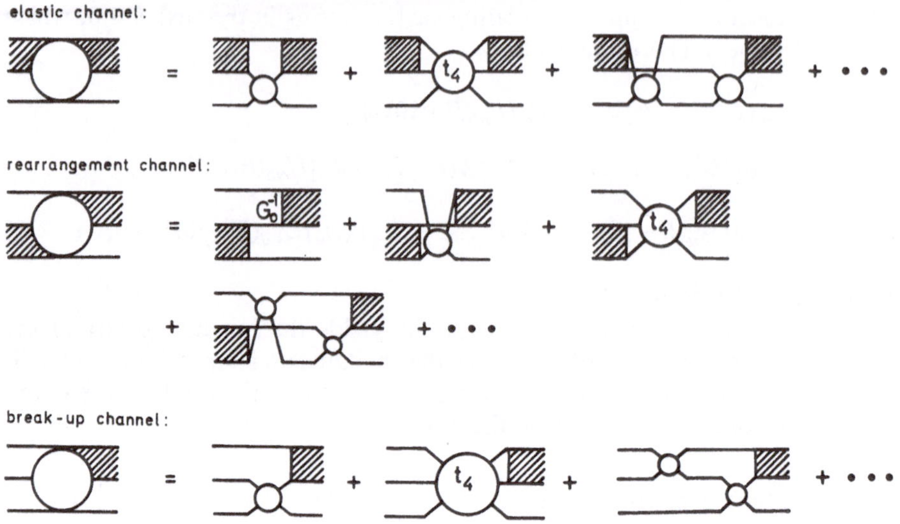

elastic channel:

rearrangement channel:

break-up channel:

Fig. 3.1. The first few terms in the multiple scattering series (3.239) for elastic, -rearrangement, and -break-up processes

How can we get insight into the convergence or divergence property of the multiple scattering series? The philosophy will be identical to the one we used for potential scattering in Sect. 1.7. There the physical reason for a divergence was the existence of bound states or resonances in the potential or the sign reversed potential. We shall find the same cause in the case of three particles.

Let us consider the eigenvalue problem

$$
G_0 \begin{pmatrix} 0 & t_1 & t_1 \\ t_2 & 0 & t_2 \\ t_3 & t_3 & 0 \end{pmatrix} \begin{pmatrix} \psi_1 \\ \psi_2 \\ \psi_3 \end{pmatrix} = \eta \begin{pmatrix} \psi_1 \\ \psi_2 \\ \psi_3 \end{pmatrix} , \tag{3.240}
$$

which generalizes the equation for the bound state (see Sect. 3.7). We have simplified the notation and dropped V_4.

What are the relevant energies? The lowest energy where scattering can take place occurs at the smallest ε_α, with the relative motion between the two fragments having zero kinetic energy. If the binding energies ε_α for the three pairs do not coincide, we have different thresholds ε_α, $\alpha = 1, 2, 3$ for the three two-body fragmentation channels. Finally at $E = 0$ the two fragments can break up into 3 particles. This basic structure of continua is shown in Fig. 3.2. The discrete value E_b stands for a possible three-body bound state energy, which has to lie below the lowest threshold for scattering of the particles.

For energies E below the thresholds, the kernel in (3.240) and the eigenvalues η are real. At $E = E_b$ there exists certainly the eigenvalue $\eta = 1$ and $|\psi_1\rangle, |\psi_2\rangle, |\psi_3\rangle$ are the Faddeev components of the bound state. In general

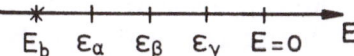

E_b ε_α ε_β ε_γ $E=0$ E

Fig. 3.2. Relevant energies for the three-body system: a discrete three-body binding energy E_b, two-body binding energies ε_α, ε_β, ε_γ, which can be different in general, and the threshold for three-body break-up $E = 0$

there exists an infinity of discrete eigenvalues. Why discrete? The reason is that the iterated kernel, $K^2(E)$, is of the Hilbert-Schmidt type for E below the thresholds. We have already mentioned that fact, which is based on the connected structure of K^2, in Sect. 3.3. Because K^2 has a discrete spectrum, K has one too. For energies below the thresholds, the Faddeev components in (3.240) decrease exponentially and are therefore square integrable. However, because of the nonlinear relation between t_α and V_α, the square integrable components do not sum up to a three body bound state of modified interactions V_α/η. Still if η becomes smaller the kernel gets stronger and one has to expect additional solutions of (3.240). Also we can expect that if V_α has attractive and repulsive parts as in nuclear physics, this freedom in sign can be felt on the level of the t-matrices and eigenvalues of both signs should occur. This is indeed the case as is known from numerical examples. Also the monotonic property of the η-values found in Sect. 1.7 carries over to three particles. If we increase the energy then the kernel has to get weaker or η larger in magnitude. Above the thresholds the components $|\psi_i\rangle$ in (3.240) will have outgoing oscillatory parts and will therefore be complex, together with η. Again the mere existence of a three-body bound state guarantees that at least one eigenvalue will be outside the unit circle for energie which are not too high. Consequently the Neumann series in the kernel K will diverge. Because of the energy denominators in the kernel, the eigenvalues will return into the unit circle, if the energy is high enough, and a perturbative treatment of the Neumann series will be justified.

Now the kernel responsible for the multiple scattering series (3.239) is a bit different from the one in (3.240). According to (3.177) the analogous eigenvalue problem would read

$$
\begin{pmatrix} 0 & t_2 & t_3 \\ t_1 & 0 & t_3 \\ t_1 & t_2 & 0 \end{pmatrix} G_0 \begin{pmatrix} |\theta_1\rangle \\ |\theta_2\rangle \\ |\theta_3\rangle \end{pmatrix} = \xi \begin{pmatrix} |\theta_1\rangle \\ |\theta_2\rangle \\ |\theta_3\rangle \end{pmatrix}. \tag{3.241}
$$

However, it is easy to relate the two problems. With the ansatz $|\theta_\alpha\rangle = G_0^{-1}(|\psi_\beta\rangle + |\psi_\gamma\rangle)$ we get

$$
G_0 t_2(|\psi_3\rangle + |\psi_1\rangle) + G_0 t_3(|\psi_1\rangle + |\psi_2\rangle) = \xi(|\psi_2\rangle + |\psi_3\rangle)
$$
$$
G_0 t_1(|\psi_2\rangle + |\psi_3\rangle) + G_0 t_3(|\psi_1\rangle + |\psi_2\rangle) = \xi(|\psi_3\rangle + |\psi_1\rangle) \tag{3.242}
$$
$$
G_0 t_1(|\psi_2\rangle + |\psi_3\rangle) + G_0 t_2(|\psi_3\rangle + |\psi_1\rangle) = \xi(|\psi_1\rangle + |\psi_2\rangle).
$$

By obvious subtractions and additions of the three equations we get back (3.240). Thus $\xi = \eta$ and the convergence or divergence criteria for the Neumann series in K carry over directly to the multiple scattering series.

One can say that in most interesting cases, the multiple scattering series will diverge because of the presence of discrete structures at neighbouring energies. Nevertheless one can extract all the necessary information as we shall outline now.

Let us first regard a method which has been made popular in few body nuclear physics by *Malfliet* and *Tjon* [3.36]. It allows us to calculate the bound state properties from the Neumann series in the kernel K of (3.240). The n'th oder term (in matrix notation) is

$$U_n = K^n U_0. \tag{3.243}$$

If we think of a decomposition of an arbitrary U_0 into components along the various eigenstates of K, then K^n produces factors η^n. As a consequence

$$\lim_{n \to \infty} \frac{U_{n+1}}{U_n} = \eta_{max}, \tag{3.244}$$

where η_{max} is the eigenvalue largest in magnitude. Now assume that at $E = E_b$, $\eta_{max} = 1$ then the recipe is obvious. One applies the kernel K n-times to an arbitrary state U_0 and stabilises the ratio in (3.244) to the desired precision. This is repeated for different energies until the ratio in (3.244) tends towards 1. At the same time, U_n will be the column vector of Faddeev components to the bound state.

In nuclear physics it can happen, that the repulsive core in a model for two-nucleon forces provides a η_{max} which is negative at $E = E_b$. An example in the case of the triton is discussed in [3.37]. Then (3.244) cannot be used as it stands and has to be modified. We can assume the decomposition

$$U_0 = \alpha_0 \chi_0 + \alpha_1 \chi_1 + \cdots, \tag{3.245}$$

where χ_0 is the eigenstate of K with the negative eigenvalue $\eta_0 = \eta_{max}$ and χ_1 is the eigenstate corresponding to the bound state, which we are looking for. Let us apply K once:

$$K U_0 = \alpha_0 \eta_0 \chi_0 + \alpha_1 \eta_1 \chi_1 + \cdots. \tag{3.246}$$

Assume that the iteration procedure (3.244) has been carried through already and we know η_0 approximately η_0^{appr}. Then we can subtract:

$$\eta_0^{appr} U_0 - K U_0 = \alpha_0 (\eta_0^{appr} - \eta_0) \chi_0 + \alpha_1 (\eta_0^{appr} - \eta_1) \chi_1 + \cdots. \tag{3.247}$$

Taking the right hand side as the new state U_0 we can regard that subtraction k-times which makes the coefficient of χ_0 as small we like: $\alpha_0(\eta_0^{appr} - \eta_0)^k$. The resulting U_0 can then be used in the Malfliet-Tjon method.

It is advantageous to combine this subtraction technique with the Padé method, which can deliver the energy E_b relatively "cheap". Consider the inhomogeneous equation

$$U(z) = U_0(z) + \lambda K(z) U(z) \tag{3.248}$$

with the kernel K defined in (3.240). Then as we saw in Sect. 2.7.3, the resolvent operator

$$F(\lambda, z) \equiv [1 - \lambda K(z)]^{-1} \tag{3.249}$$

has poles at $\lambda = 1/\eta$. Note that the poles $\lambda_0 = 1/\eta_0$ and $\lambda_1 = 1/\eta_1$ from the example above are different, and at $z = E_b$ the actual strength parameter $\lambda = 1$ coincides with the pole λ_1. The Padé ratio provides an approximation to the resolvent operator. Therefore one has simply to determine the energy at which the inverse of the Padé ratio goes to zero. That energy is E_b. The Padé ratio itself is the desired physical eigenstate of K (up to a large normalisation factor). This is indeed a means of determining the eigenstate itself [3.38, 39] which requires, however, the calculation of the Padé ratio for *all* momentum variables, whereas the energy search can be done just for an arbitrary and fixed choice of momentum variables. Therefore an alternative and maybe "cheaper" method is to use the modified Malfliet-Tjon technique once E_b is determined.

Let us finally return to the scattering problem which is the subject of this chapter. We can be very brief. The multiple scattering series can be used directly as the input for the Padé method. A very enlightning study on the multiple scattering series for a nuclear three-body model and its Padé summation has been presented in [3.40]. We shall regard examples in Sect. 3.5.3.

3.4.4 Identical Particles

In nuclear physics, one can treat neutrons and protons as identical particles by adding the isospin quantum numbers. This leads to a formal simplification. We shall regard the AGS-equations (3.211) for the transition operators, the expressions (3.212) and (3.213) for the break-up operator, and the unitarity relations (3.238).

The states have to be symmetrized in the case of bosons and antisymmetrized in the case of fermions. In the following we shall simply talk of symmetrizations and mean both types of statistics. Let us regard a state with two fragments in the initial channel. The channel state $|\phi_\alpha\rangle = |\varphi_\alpha\rangle |q_\alpha\rangle$ can be assumed to be already symmetrized in the two body subsystem α. In other words, $|\varphi_\alpha\rangle$ is supposed to have that property. Then the symmetrized scattering state is

$$|\Psi_\mathscr{S}\rangle = \mathscr{S} \lim_{\varepsilon\to 0} i\varepsilon G |\phi_1\rangle \tag{3.250}$$

with

$$\mathscr{S} = 1 + P_{12}P_{23} + P_{13}P_{23} \equiv 1 + P. \tag{3.251}$$

Since $\mathscr{S}$ commutes with the symmetric operator G, $\mathscr{S}$ acts directly on $|\phi_1\rangle$, and we get

$$\mathscr{S}|\phi_1\rangle = |\phi_1\rangle + |\phi_2\rangle + |\phi_3\rangle. \tag{3.252}$$

Note that $|\phi_2\rangle$ and $|\phi_3\rangle$ carry the same quantum numbers (momenta, spins etc.) as $|\phi_1\rangle$, but refer to different particles. A simple inspection reveals that the rhs of (3.252) is totally antisymmetric or symmetric, depending on the types of particles. Inserting (3.252) into (3.250) yields the symmetrized state

$$|\Psi_\mathscr{S}\rangle = |\Psi_1^{(+)}\rangle + |\Psi_2^{(+)}\rangle + |\Psi_3^{(+)}\rangle. \tag{3.253}$$

Obviously the three initial two-fragment configurations are treated completely democratically, and are not distinguishable any more. Of course the same $|\Psi_\mathscr{S}\rangle$ would have resulted if $|\phi_2\rangle$ or $|\phi_3\rangle$ would have been chosen in (3.250).

What will the elastic scattering amplitude be now? We consider the situation in the final state that one particle is far away from a bound pair of the other two. Let us call it particle 1. Then the outgoing flux carries the amplitude

$$\begin{aligned}(\phi_1|V^1|\Psi_\mathscr{S}\rangle &= \langle\phi_1|P_{23}P_{12}P_{12}P_{23}V^1|\Psi_\mathscr{S}\rangle \\ &= \langle\phi_2|V^2|\Psi_\mathscr{S}\rangle = \langle\phi_3|V^3|\Psi_\mathscr{S}\rangle.\end{aligned} \tag{3.254}$$

In the second equality we applied the cyclical permutation to the left and right and used the invariance of $|\Psi_\mathscr{S}\rangle$. The last equality follows upon insertion of $P_{23}P_{13}P_{13}P_{23}$. Thus the outgoing fluxes carry of course the same amplitudes, whether particle 1 is far away from (2 3) or 2 from (31) or 3 from (1 2). Note that the quantum numbers of the channel states in (3.245) are all the same; the indices suggest only three different choices of integration variables. We can now introduce the single transition operator U by

$$\langle\phi_1|V^1|\Psi_\mathscr{S}\rangle = \sum_{\alpha=1}^{3} \langle\phi_1|V^1|\Psi_\alpha^{(+)}\rangle = \sum_{\alpha=1}^{3} \langle\phi_1|U_{1\alpha}|\phi_\alpha\rangle \equiv \langle\phi_1|U|\phi_1\rangle. \tag{3.255}$$

Simply related to $U|\phi_1\rangle$ is $V^2|\Psi_{\mathscr{S}}\rangle$ and $V^3|\Psi_{\mathscr{S}}\rangle$. Namely

$$V^2|\Psi_{\mathscr{S}}\rangle = \sum_\alpha U_{2\alpha}|\phi_\alpha\rangle = P_{12}P_{23}V^1|\Psi_{\mathscr{S}}\rangle$$
$$= P_{12}P_{23}\sum U_{1\alpha}|\phi_\alpha\rangle = P_{12}P_{23}U|\phi_1\rangle \qquad (3.256)$$

and

$$V^3|\Psi_{\mathscr{S}}\rangle = \sum_\alpha U_{3\alpha}|\phi_\alpha\rangle = P_{13}P_{23}U|\phi_1\rangle . \qquad (3.257)$$

Now we are prepared to derive the integral equation for U. Let us first regard (3.164) for instance for $\beta = 1$. We sum over α and get

$$\sum_\alpha U_{1\alpha}|\phi_\alpha\rangle = \sum_\alpha \bar{\delta}_{1\alpha}G_0^{-1}|\phi_\alpha\rangle + \sum_{\gamma \neq 1} t_\gamma G_0 \sum_\alpha U_{\gamma\alpha}|\phi_\alpha\rangle . \qquad (3.258)$$

According to (3.255, 256) and (3.257) this can be fully expressed in terms of the U-operator:

$$U|\phi_1\rangle = PG_0^{-1}|\phi_1\rangle + Pt_1G_0U|\phi_1\rangle , \qquad (3.259)$$

where P is defined in (3.251). Though we end up with one equation instead of three, the full complexity is still present, namely the recoupling requirement contained now in the permutation operators. Of course one can drop the index 1 which only reminds us of a specific choice of variables.

If we include the three-body force, we can use for instance (3.211) and repeating the steps find

$$U|\phi_1\rangle = PG_0^{-1}|\phi_1\rangle + (1+P)t_4|\phi_1\rangle$$
$$+ Pt_1G_0U|\phi_1\rangle + (1+P)t_4G_0t_1G_0U|\phi_1\rangle . \qquad (3.260)$$

According to present day insight, the three-nucleon force appears to be weak in comparison to two-nucleon forces and a perturbative treatment may be justified. As an example we shall regard the lowest order approximation:

$$t_4 \approx V_4 . \qquad (3.261)$$

Moreover we write the three-body force as a sum over three cyclical terms

$$V_4 = V_4^{(1)} + V_4^{(2)} + V_4^{(3)} = V_4^{(1)} + P_{12}P_{23}V_4^{(1)}P_{13}P_{23} + P_{13}P_{23}V_4^{(1)}P_{12}P_{23} . \qquad (3.262)$$

This force is totally symmetric, as it should be if $V_4^{(1)}$ is symmetric under exchange of particles 2 and 3. A popular model for a three-nucleon force [3.41, 42] leads to this decomposition quite naturally. It is the force resulting

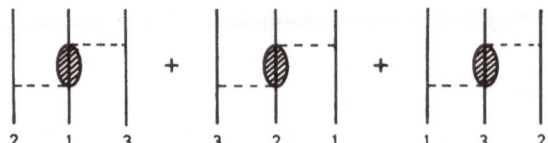

Fig. 3.3. The two-pion-exchange three-nucleon force

from a two-pion exchange between three nucleons as shwon in Fig. 3.3. The blob represents the general π-N off-shell scattering amplitude, where the intermediate nucleon state (the so called forward propagating Born term) is subtracted out. That intermediate nucleon state clearly generates the free nucleon propagator G_0 and the diagrams would just describe the action of pair forces and not a three-nucleon force. For instance, the first diagram in Fig. 3.3 with an intermediate nucleon state would be $V_{21}^{\pi} G_0 V_{13}^{\pi} + V_{13}^{\pi} G_0 V_{21}^{\pi}$, where V_{ij}^{π} is the one-pion exchange potential between two nucleons, and is part of the two-nucleon force. This iteration of pair interactions is included automatically if one solves the Schrödinger equation with pair interactions.

The three diagrams in Fig. 3.3 each singles out one of particle 1, 2, or 3 but are otherwise identical, and turn into each other through cyclical permutations. Also crossing symmetry for the π-N amplitude generates the required symmetry between particles 2 and 3 in the first diagram of Fig. 3.3 for instance.

Now we can use the property (3.262) to find

$$(1+P) V_4 = (1+P) V_4^{(1)} (1+P) \tag{3.263}$$

and (3.260), with the approximation (3.261), turns into

$$U = PG_0^{-1} + (1+P) V_4^{(1)} (1+P) + PtG_0 U + (1+P) V_4^{(1)} (1+P) G_0 t U. \tag{3.264}$$

We tacitly assumed that both sides should be applied to ϕ and P should create the two cyclical permutations of the arrangement defined by ϕ.

We now regard the break-up operator. According to (3.212) the break-up amplitude is

$$\sum_\alpha \langle \phi_0 | U_{0\alpha} | \phi_\alpha \rangle = \langle \phi_0 | t_4 (1+P) | \phi_1 \rangle + \langle \phi_0 | (1 + t_4 G_0)(1+P) t G_0 U | \phi_1 \rangle$$

$$\equiv \langle \phi_0 | U_0 | \phi_1 \rangle \tag{3.265}$$

or

$$U_0 = (1+P) t_4 + (1+P)(1 + t_4 G_0) t G_0 U. \tag{3.266}$$

From a numerical point of view it may be advantageous to evaluate U_0 in a form suggested by (3.266):

$$U_0 = (1+P) U_0^{(1)} = U_0^{(1)} + U_0^{(2)} + U_0^{(3)}. \tag{3.267}$$

Here the terms $U_0^{(2)}$ and $U_0^{(3)}$ have the same functional form as $U_0^{(1)}$ but depend on cyclically permuted variables.

Finally comparing (3.266) with (3.260) on-shell, reveals the form

$$U_0 = (1 + t G_0) U, \tag{3.268}$$

which also follows of course from (3.213) after symmetrization.

We conclude this section by regarding the unitarity relation obeyed by U and U_0. We call the two-channel states ϕ_α and ϕ_β, which are different in general, ϕ and ϕ'. Let us then sum (3.238) over α and β. The procedure will be clear through the following example:

$$\sum_{\alpha,\beta=1}^{3} \langle \phi_\alpha | U_{\alpha\beta}(E+i0) | \phi_\beta \rangle = \sum_\alpha \langle \phi_\alpha | U | \phi' \rangle = 3 \langle \phi | U | \phi' \rangle. \tag{3.269}$$

It follows that

$$\langle \phi' | U | \phi \rangle^* - \langle \phi | U | \phi' \rangle = \int dq \, \langle \phi_q | U | \phi \rangle^* \, 2\pi i \, \delta(E - E_q) \langle \phi_q | U | \phi' \rangle$$
$$+ \tfrac{1}{6} \int dp \, dq \, \langle \phi_0^{\mathscr{S}} | U_0 | \phi \rangle^* \, 2\pi i \, \delta(E - E_{pq}) \langle \phi_0^{\mathscr{S}} | U_0 | \phi' \rangle. \tag{3.270}$$

Here we have introduced the free state

$$| \phi_0^{\mathscr{S}} \rangle \equiv \frac{1}{\sqrt{2}} (1 \pm P_{23}) | \phi_0 \rangle,$$

which is symmetrized in the pair (2 3). For $| \phi \rangle = | \phi' \rangle$ one recovers the optical theorem. In that case, the second term on the rhs is the total break-up cross section and the factor 1/6 removes the 6-fold overcounting caused by the p-q-integration.

Supplement. It is interesting to regard the set of Lippmann-Schwinger equation (3.51 a – c) in the case of identical particles. According to (3.253) the symmetrized state is

$$| \Psi^{\mathscr{S}} \rangle = | \Psi_1 \rangle + | \Psi_2 \rangle + | \Psi_3 \rangle. \tag{S.1}$$

Let us now indicate explicitly the dependence on the variables of the three particles. As a short hand notation for states, where space-, spin-, isospin-, and possibles further variables for the three particles have fixed values, we introduce

$$| 1\,2\,3 \rangle_1 = | 2\,3\,1 \rangle_2 = | 3\,1\,2 \rangle_3. \tag{S.2}$$

In each case particles 1, 2, and 3 are at the "positions 1, 2, and 3", respectively. Then a representation is defined as

$$\Psi_1(1\ 2\ 3) \equiv {}_1\langle 1\ 2\ 3|\Psi_1\rangle = {}_2\langle 2\ 3\ 1|\Psi_1\rangle = {}_3\langle 3\ 1\ 2|\Psi_1\rangle \tag{S.3}$$

For identical particles clearly the state $|\Psi_2\rangle$ initiated through channel 2 is identical to the state $|\Psi_1\rangle$, if particles are suitably permuted. Namely one has

$$|\Psi_2\rangle = P_{12}P_{23}|\Psi_1\rangle \tag{S.4}$$

and similarily

$$|\Psi_3\rangle = P_{13}P_{23}|\Psi_1\rangle . \tag{S.5}$$

This reads in our representation (S.2)

$$\Psi_2(1\ 2\ 3) \equiv {}_1\langle 1\ 2\ 3|\Psi_2\rangle$$
$$= {}_1\langle 1\ 2\ 3|P_{12}P_{23}|\Psi_1\rangle$$
$$= {}_3\langle 1\ 2\ 3|\Psi_1\rangle = {}_1\langle 2\ 3\ 1|\Psi_1\rangle = \Psi_1(2\ 3\ 1) \tag{S.6}$$

and

$$\Psi_3(1\ 2\ 3) = \Psi_1(3\ 1\ 2) . \tag{S.7}$$

In other words, for a certain choice of quantum numbers in the initial state, there is only one state

$$\Psi(1\ 2\ 3) \equiv \lim_{\varepsilon \to 0} i\varepsilon\, G(E + i\varepsilon)\, \phi(1\ 2\ 3) \tag{S.8}$$

and the symmetrized state is

$$\Psi^{\mathscr{S}}(1\ 2\ 3) = \Psi(1\ 2\ 3) + \Psi(2\ 3\ 1) + \Psi(3\ 1\ 2) . \tag{S.9}$$

What are the Lippmann-Schwinger equations which determine $\Psi^{\mathscr{S}}(1\ 2\ 3)$ uniquely? The answer is obvious. One needs the following set

$$\Psi^{\mathscr{S}}(1\ 2\ 3) = \phi(1\ 2\ 3) + G_1 V^1 \Psi^{\mathscr{S}}(1\ 2\ 3)$$
$$\Psi^{\mathscr{S}}(1\ 2\ 3) = \phi(2\ 3\ 1) + G_2 V^2 \Psi^{\mathscr{S}}(1\ 2\ 3) \tag{S.10}$$
$$\Psi^{\mathscr{S}}(1\ 2\ 3) = \phi(3\ 1\ 2) + G_3 V^3 \Psi^{\mathscr{S}}(1\ 2\ 3) .$$

The first equation alone would allow the solution $\Psi(1\ 2\ 3) + c_2\Psi(2\ 3\ 1) + c_3\Psi(3\ 1\ 2)$, where c_2 and c_3 are arbitrary. The constants are fixed to unity by requiring the two additional equations in (S.10).

One can go one step further. In [3.6] we proposed one equation which defines $|\Psi^{\mathscr{S}}\rangle$, namely

$$\Psi^{\mathscr{S}}(1\,2\,3) = \phi(1\,2\,3) + G_1 V_2 \Psi^{\mathscr{S}}(2\,3\,1) + G_1 V_3 \Psi^{\mathscr{S}}(3\,1\,2) \,. \tag{S.11}$$

Interestingly enough this is just the coupling scheme rediscovered later and called the Faddeev-Lovelace choice, that we described in Sect. 3.3.3. We may take over the proof of that section to demonstrate that the only solution of (S.11) is just (S.9). We operate by $(1 - G_0 V_1)$ from the left and get

$$\Psi^{\mathscr{S}}(1\,2\,3) = G_0 V_1 \Psi^{\mathscr{S}}(1\,2\,3) + G_0 V_2 \Psi^{\mathscr{S}}(2\,3\,1) + G_0 V_3 \Psi^{\mathscr{S}}(3\,1\,2) \,. \tag{S.12}$$

The rhs is totally symmetrized, which implies the following property of every solution of (S.11)

$$\Psi^{\mathscr{S}}(1\,2\,3) = \Psi^{\mathscr{S}}(2\,3\,1) = \Psi^{\mathscr{S}}(3\,1\,2) \tag{S.13}$$

Consequently (S.11) reverts back to the three equations (S.10). The problem formulated in (S.11) has been solved approximately in a simple three-body model [3.24]. The solution exhibits the symmetry property (S.13) automatically as it should, of course.

3.5 Examples of Numerical Studies in Few-Nucleon Scattering

Much work has been devoted to solving the Faddeev equations in the scattering region. We shall indicate a few approaches and mention a few numerical studies. This is a very subjective selection and we have to apologize to the authors, whose work is not directly mentioned. We emphasize this point, since every honest solution of this problem requires very hard work. One faces a problem with two vector variables and in addition spin- and isospin degrees of freedom. Present day standard numerical techniques have to be pushed to their very limits to achieve converged solutions.

3.5.1 Lovelace Equations

We remember from Sect. 1.5 that the two-body t-matrix has a pole at a bound state energy and the residue factorizes:

$$\hat{t}(z) \xrightarrow[z \to \varepsilon_b]{} \frac{V|\varphi_b\rangle\langle\varphi_b|V}{z - \varepsilon_b} \,. \tag{3.271}$$

In other words $\hat{t}$ is a finite rank operator. A closer inspection shows that this is also true for z at resonance energies on the second sheet. This leads to an essential technical simplification and an important insight into the physical mechanism of a three-body scattering process. This step has been pioneered by *Lovelace* [3.16].

To present the basic structure, let us first assume that the $\hat{t}$-operator in the two-body space has the form

$$\hat{t}(z) = |g\rangle\,\tau(z)\,\langle g| \tag{3.272}$$

not only in the immediate neighbourhood of the poles but everywhere. We shall comment later on how systematic corrections can be included to this type of approximation.

Of course (3.272) is not an approximation if the underlying two-body potential is already separable:

$$V = \lambda\,|g\rangle\langle g| . \tag{3.273}$$

Then the Lippmann-Schwinger equation (1.128) for $\hat{t}(z)$ can obviously be solved algebraically with a result of the form (3.272).

Exercise: Determine $\hat{t}(z)$ for the pair interaction (3.273).

In case the so called form factor $|g\rangle$ in (3.273) depends on z, it has to have the limiting property

$$|g\rangle \to V|\varphi_b\rangle \quad \text{for} \quad z \to \varepsilon_b . \tag{3.274}$$

In the three-body space (3.272) reads

$$\langle p'q'|t(z)|pq\rangle = \delta^3(q-q')\langle p'|\hat{t}(z - \tfrac{3}{4}q^2)|p\rangle$$
$$= \delta^3(q-q')g(p')\,\tau(z - \tfrac{3}{4}q^2)g(p) . \tag{3.275}$$

This separable structure in p and p' allows us to reduce the two-vector-variable integral equations to one variable ones. Let us insert into the AGS-equations (3.164)

$$U_{\alpha\beta} = \bar{\delta}_{\alpha\beta}G_0^{-1} + \sum_{\gamma \neq \alpha} t_\gamma G_0 U_{\gamma\beta} \tag{3.276}$$

the separable form (3.272):

$$U_{\alpha\beta} = \bar{\delta}_{\alpha\beta}G_0^{-1} + \sum_{\gamma \neq \alpha} |g_\gamma\rangle\,\tau_\gamma\langle g_\gamma|G_0 U_{\gamma\beta} . \tag{3.277}$$

Note that the second term on the right hand side already factorizes the p- and q-dependence:

$$\sum_{\gamma \neq \alpha} \langle p | g_\gamma \rangle \tau_\gamma (z - \tfrac{3}{4} q^2) \langle q | \langle g_\gamma | G_0 U_{\gamma\beta}. \tag{3.278}$$

Regarding the right hand side, the unknown operators occur only in the form $\langle g_\gamma | G_0 U_{\gamma\beta}$, which leads us necessarily to create that term also on the lhs:

$$\langle g_\alpha | G_0 U_{\alpha\beta} = \bar{\delta}_{\alpha\beta} \langle g_\alpha | G_0 G_0^{-1} + \sum_{\gamma \neq \alpha} \langle g_\alpha | G_0 | g_\gamma \rangle \tau_\gamma \langle g_\gamma | G_0 U_{\gamma\beta}. \tag{3.279}$$

Now all the operators will be applied to the channel state $| \phi_\beta \rangle$, which obeys

$$G_0(E) V_\beta | \phi_\beta \rangle = | \phi_\beta \rangle \tag{3.280}$$

for $E = \varepsilon_\beta + \tfrac{3}{4} q_\beta^2$ (on shell). This can also be written, due to (3.274), as

$$| \phi_\beta \rangle = G_0 | g_\beta \rangle | q_\beta \rangle \big|_{\text{on shell}}. \tag{3.281}$$

Therefore it is clear that we should apply (3.279) onto $G_0 | g_\beta \rangle$, and one gets the nice symmetric form

$$\langle g_\alpha | G_0 U_{\alpha\beta} G_0 | g_\beta \rangle = \bar{\delta}_{\alpha\beta} \langle g_\alpha | G_0 | g_\beta \rangle$$
$$+ \sum_{\gamma \neq \alpha} \langle g_\alpha | G_0 | g_\gamma \rangle \tau_\gamma \langle g_\gamma | G_0 U_{\gamma\beta} G_0 | g_\beta \rangle. \tag{3.282}$$

The individual terms are still operators in the space of relative momentum states $| q_\lambda \rangle$. Let us call them $X_{\alpha\beta}$ and $Z_{\alpha\beta}$. Then (3.282) reads

$$\langle q_\alpha | X_{\alpha\beta} | q_\beta \rangle = \langle q_\alpha | Z_{\alpha\beta} | q_\beta \rangle$$
$$+ \sum_{\gamma \neq \alpha} \int dq_\gamma \langle q_\alpha | Z_{\alpha\gamma} | q_\gamma \rangle \tau_\gamma (z - \tfrac{3}{4} q_\gamma^2) \langle q_\gamma | X_{\gamma\beta} | q_\beta \rangle, \tag{3.283}$$

which has the structure of multichannel, two-body, Lippmann-Schwinger equations.

The quantity τ plays the role of the free propagator of particle γ with respect to the pair "γ". Indeed τ_γ has a pole at the two-body binding energy:

$$\tau(z - \tfrac{3}{4} q_\gamma^2) \approx (z - \tfrac{3}{4} q_\gamma^2 - \varepsilon_\gamma)^{-1} \text{ near the pole}. \tag{3.284}$$

Further, $\langle q_\alpha | Z_{\alpha\gamma} | q_\gamma \rangle$ acts as a transition potential between different channels. For on shell momenta this is very apparent:

$$\bar{\delta}_{\alpha\beta} \langle q_\alpha | \langle g_\alpha | G_0 | g_\beta \rangle | q_\beta \rangle = \bar{\delta}_{\alpha\beta} \langle \phi_\alpha | V_\alpha | \phi_\beta \rangle. \tag{3.285}$$

The potential Z is nonlocal, which is not surprising for composite particle scattering. Finally, the physical transition amplitudes are given through the on-shell values

$$\langle q_\alpha | \langle g_\alpha | G_0 U_{\alpha\beta} G_0 | g_\beta \rangle | q_\beta \rangle |_{\text{on shell}} = \langle \phi_\alpha | U_{\alpha\beta} | \phi_\beta \rangle . \tag{3.286}$$

In this approximation, the break-up process also achieves a simple picture. According to (3.168), and using (3.272), one gets

$$\langle pq | U_{0\alpha} | \phi_\alpha \rangle = \sum_\gamma \langle p_\gamma | g_\gamma \rangle \, \tau_\gamma (z - \tfrac{3}{4} q_\gamma^2) \langle q_\gamma | \langle g_\gamma | G_0 U_{\gamma\alpha} G_0 | g_\alpha \rangle | q_\alpha \rangle . \tag{3.287}$$

Thus the break-up occurs via a scattering into all two fragment channels, and only then the break-up of the pairs follows.

This appealing picture, proposed by Lovelace, provoked of course the question of to what extent a general t-matrix for local (nonseparable) forces can be approximated by finite rank operators. We mentioned one possibility in Sect. 1.7, which is based on a finite rank approximation of the kernel in the Lippmann-Schwinger equation for $\hat{t}$. At the same time one would like a solution which satisfies unitarity. A very successful and popular approximation that achieves this is called the unitary pole approximation, UPA [3.43]. Another one, which is similar in quality (if not better), is the Adhikari-Sloan expansion [3.44]. We refer the interested reader to the original literature for more information.

Once a good approximation of finite rank, $\hat{t}_s$ (in general several terms), for $\hat{t}$ is found, it remains to incorporate the rest $\hat{t}'$ in a systematic manner. This has been worked out in a very transparent way for the AGS-equations [3.31]. The technique is essentially the same as described in Sect. 1.7. Presenting $\hat{t}$ as a sum over $\hat{t}_s$ plus $\hat{t}'$ the AGS-equation (3.276) read

$$U_{\alpha\beta} = \bar{\delta}_{\alpha\beta} G_0^{-1} + \sum_{\gamma \neq \alpha} t_\gamma' G_0 U_{\gamma\beta} + \sum_\gamma \bar{\delta}_{\gamma\alpha} G_0^{-1} G_0 t_\gamma^s G_0 U_{\gamma\beta} . \tag{3.288}$$

One defines solutions $U'_{\alpha\beta}$ based on t_γ' alone:

$$U'_{\alpha\beta} = \bar{\delta}_{\alpha\beta} G_0^{-1} + \sum_{\gamma \neq \alpha} t_\gamma' G_0 U'_{\gamma\beta} . \tag{3.289}$$

Since t_γ' is assumed to be small, this system can be solved by iteration. In other words, the lowest orders should be sufficient. Then (3.288) can obviously be rewritten as

$$U_{\alpha\beta} = U'_{\alpha\beta} + \sum_\gamma U'_{\alpha\gamma} G_0 | g_\gamma \rangle \, \tau_\gamma \langle g_\gamma | G_0 U_{\gamma\beta} . \tag{3.290}$$

As an example we have inserted t_{γ}^s as a one-term separable form. We recognize the same structure as in (3.277) and repeating the steps leading to (3.283) we arrive at

$$X_{\alpha\beta} = Z'_{\alpha\beta} + \sum_{\gamma} Z'_{\alpha\gamma} \tau_{\gamma} X_{\gamma\beta} \tag{3.291}$$

with

$$Z'_{\alpha\beta} = \langle g_{\alpha}|G_0 U'_{\alpha\beta} G_0|g_{\beta}\rangle . \tag{3.292}$$

The lowest order approximation of $U'_{\alpha\beta}$ gives back the potential $Z_{\alpha\beta}$ based on a purely separable $\hat{t}$-matrix.

3.5.2 Kinematical Curves [3.45]

In the break-up configuration, the energy can be continuously distributed over the two relative motions. Besides energy conservation, momentum conservation imposes constraints on the accessible break-up states. Let us first work in the center-of-mass system where the total three momentum is zero. Then the conservation laws read

$$\sum_i k_{i,c} = 0 \tag{3.293}$$

$$\sum_i \frac{k_{i,c}^2}{2m_i} \equiv \sum_i E_i^c = E_{CM} . \tag{3.294}$$

Inserting $k_{3,c}$ from (3.293) into (3.294) yields immediately

$$E_1^c(1 + m_1/m_3) + E_2^c(1 + m_2/m_3) + \frac{2\sqrt{E_1^c E_2^c m_1 m_2}}{m_3} \cos \vartheta_{12}^c = E_{CM} . \tag{3.295}$$

Here ϑ_{12}^c is the angle between $k_{1,c}$ and $k_{2,c}$. Equation (3.295) describes an ellipse in the variables $\sqrt{E_1^c}$, $\sqrt{E_2^c}$. The kinematically allowed events have to lie on that curve. An example for $\vartheta_{12}^c = 60°$, $E_{CM} = 33.33$ MeV and equal mass particles is shown in Fig. 3.4a, b for the variables $\sqrt{E_i^c}$ and E_i^c, respectively. Note that only part of the closed curves are accessible as real events.

If one studies final state interactions between pairs of particles, one likes to know the points on the kinematical curve which belong to a fixed relative energy within that pair. In nuclear physics for instance, the t-matrix in the $^1S_0$-state is enhanced near zero energy because of the anti-bound state pole (see Sect. 1.6). Therefore one has to expect, that the final-state interaction in that state is especially strong if the relative energy for that pair goes to zero. A famous example is the determination of the low energy parameters in the two-

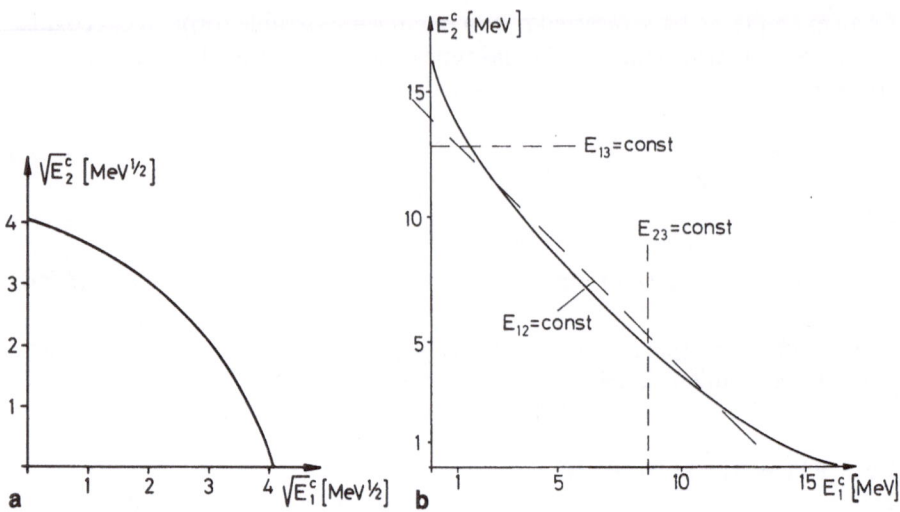

Fig. 3.4. (a) An example for a kinematical curve in the center-of-mass system in the variables $\sqrt{E_i^c}$. **(b)** The same curve as in **(a)** in the variables E_i^c. Curves for constant relative pair energies are indicated by dotted lines

neutron system, scattering length and effective range, through their final-state interaction in the reaction $n + d \rightarrow n + n + p$ [3.46] at low relative energies.

We know from Sect. 3.1 that the total energy E_{CM} is the sum of the two energies of relative motion. Generalizing to arbitrary masses (3.4) reads

$$E_{CM} = \frac{p_1^2}{2\mu_{23}} + \frac{q_1^2}{2M_{23}}$$

(3.296)

with

$$\frac{1}{\mu_{23}} = \frac{1}{m_2} + \frac{1}{m_3}$$

(3.297)

$$\frac{1}{M_{23}} = \frac{1}{m_1} + \frac{1}{m_2 + m_3} \,.$$

(3.298)

Since $q_1 = k_{1,c}$ in the center-of-mass system we get

$$E_{CM} = E_{23} + E_1^c \frac{m_1}{M_{23}} = E_{23} + E_1^c \frac{m_1 + m_2 + m_3}{m_2 + m_3} \,.$$

(3.299)

Thus, fixing the value E_1^c is equivalent to specifying the relative energy in the pair (2 3). Specifically, the kinematically allowed, minimal value of E_{23} occurs on the curve where E_1^c is maximal. Similarily the conditions for constant relative energies in the pairs (1 3) and (1 2) are given by

$$E_{CM} = E_{13} + E_2^c \frac{m_1 + m_2 + m_3}{m_3 + m_1} \tag{3.300}$$

$$E_{CM} = E_{12} + (E_{CM} - E_1^c - E_2^c) \frac{m_1 + m_2 + m_3}{m_1 + m_2} . \tag{3.301}$$

The three straight lines (3.299, 300) and (3.301) are indicated in Fig. 3.4b as dotted lines.

The kinematically allowed curves in the laboratory system are equally easy to achieve. Let P be the projectile momentum and E_p^l its energy. By definition of the laboratory system the target (t) is rest. Therefore the conservation laws read

$$\sum k_{i,l} = P \tag{3.302}$$

$$\sum \frac{k_{i,l}^2}{2m_i} \equiv \sum_i E_i^l = E_p^l + Q \tag{3.303}$$

with

$$Q = m_t + m_p - m_1 - m_2 - m_3 . \tag{3.304}$$

In our examples, $Q = \varepsilon < 0$ is the bound state energy of the pair in the initial state. Again eliminating $k_{3,l}$ one finds

$$\frac{1}{m_3} [E_1^l(m_1 + m_3) + E_2^l(m_2 + m_3)$$

$$- 2\sqrt{m_p m_1 E_p^l E_1^l} \cos \vartheta_1^l - 2\sqrt{m_p m_2 E_p^l E_2^l} \cos \vartheta_2^l$$

$$+ 2\sqrt{m_1 m_2 E_1^l E_2^l} \cos \vartheta_{12}^l] = Q + E_p^l (1 - m_p/m_3) \tag{3.305}$$

with

$$\cos \vartheta_i^l \equiv \hat{k}_{i,l} \hat{P} \tag{3.306}$$

and

$$\cos \vartheta_{12}^l \equiv \hat{k}_1 \hat{k}_2 . \tag{3.307}$$

This is again an ellipse in the variables $\sqrt{E_1^l}$ and $\sqrt{E_2^l}$. An example is shown in the variables $\sqrt{E_1^l}$ and E_1^l in Fig. 3.5 a, b.

Now where are the lines of constant relative energies E_{ij}? Since, for example, a constant E_{23} is equivalent to a constant E_1^c as we saw in (3.299), we need the connection between E_1^c and E_1^l. This requires a short remark. In a transition from a two fragment channel to a break-up three-body channel one

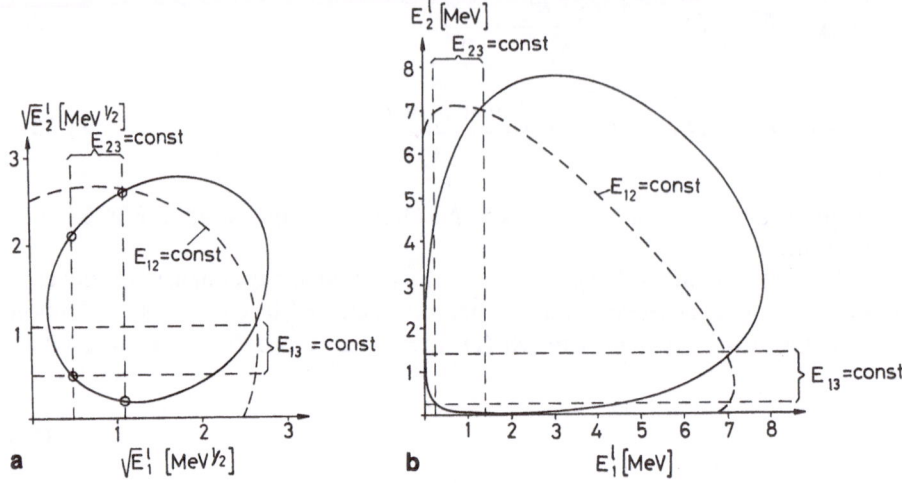

Fig. 3.5. (a) An example for a kinematical curve in the laboratory system in the variables $\sqrt{E_i^1}$.
(b) The same curve as in Fig. 3.5(a) in the variables E_i^1. Curves for constant relative pair energies
are indicated by dotted lines

has to pay some attention to the nonrelativistic form of the center-of-mass
velocity. Its general form is

$$V = \frac{P}{P_{0,l}}, \tag{3.308}$$

where P is the total three momentum and $P_{0,l}$ the total energy in the laboratory system. Because of four momentum conservation, V clearly remains the
same for the initial and final channel. The total laboratory energy is

$$P_{0,l} = m_t + \sqrt{m_p^2 + P^2} = \sqrt{m_1^2 + k_{1,l}^2} + \sqrt{m_2^2 + k_{2,l}^2} + \sqrt{m_3^2 + k_{3,l}^2}. \tag{3.309}$$

In a nonrelativistic reduction, this yields

$$P_{0,l} = m_t + m_p + E_p' = \sum m_i + \sum E_i', \tag{3.310}$$

where the kinetic energies E_j' have the standard nonrelativistic form. Therefore one may write V as

$$V \approx \frac{P}{m_t + m_p} \quad \text{or} \quad V \approx \frac{P}{\sum m_i}, \tag{3.311}$$

which are only equal if the binding energy Q can be neglected with respect to
the rest masses. To the accuracy required by nuclear physics this is usually

the case. Let us take the second form. Then the Galilean transformation to the system where V or P is zero is simply

$$k_{i,c} = k_{i,l} - \frac{m_i}{\sum m_j} P . \tag{3.312}$$

This leads immediately to the connection between lab- and center-of-mass energies:

$$E_i^c = (\sqrt{E_i^l} - a_i \cos \vartheta_i^l)^2 + a_i^2 \sin^2 \vartheta_i^l \tag{3.313}$$

with

$$a_i = \frac{2\sqrt{m_i m_p E_p^l}}{\sum m_j} \tag{3.314}$$

and

$$\cos \vartheta_i^l = \hat{P}_{i,l} \hat{P} . \tag{3.315}$$

We recognize that, in general, there are two laboratory energies E_i^l which lead to the same E_i^c or E_{jk}. This is indicated by dotted lines in Fig. 3.5 a, b. To draw the curves for fixed E_3^c on the figure spanned by E_1^l and E_2^l, one expresses E_3^c in terms of E_1^c and E_2^c and gets

$$E_3^c = E_{CM} - E_1^c - E_2^c = E_{CM} - a_1^2 \sin^2 \vartheta_1^l$$
$$- a_2^2 \sin^2 \vartheta_2^l - (\sqrt{E_1^l} - a_1 \cos \vartheta_1^l)^2 - (\sqrt{E_2^l} - a_2 \cos \vartheta_2^l)^2 . \tag{3.316}$$

This is a circle in $\sqrt{E_1^l}$, $\sqrt{E_2^l}$ space and intersects the ellipse (3.305) at most 4 times.

We want to illustrate a kinematical situation where the three final momenta lie in a plane containing the initial beam. Also we choose three equal masses, $E_p^l = 50$ MeV and $Q = 4$ MeV. The Fig. 3.5 a, b correspond to $\vartheta_1^l = 70°$, $\vartheta_2^l = -70°$ and $\vartheta_{12} = 140°$. Specifically the intersections generated by the right dotted vertical line ($E_1^c = 5$ MeV) in Fig. 3.5 a, b belong to the two cases shown in Fig. 3.6. Inserted are also the center-of-mass momenta $k_{1,c}$ and $k_{2,c}$ as given through (3.312).

Note that a kinematical curve for the center-of-mass system, as shown in Fig. 3.4 a, b belongs to a fixed angle between $k_{1,c}$ and $k_{2,c}$. Therefore the lab- and center-of-mass curves are not mapped onto each other. Changing the length of $k_{1,l}$ and $k_{2,l}$ but keeping their directions fixed maps the lab-curve but changes ϑ_{12}^c as is obvious from the two cases shown in Fig. 3.6.

The second possibility for E_1^l allowed by (3.313) is shown in Fig. 3.7. This corresponds to the left dotted vertical line in Fig. 3.5 a, b. Though the 4 cases, indicated in Fig. 3.5 a by circles, yield the same value of E_1^c and consequently the same relative energies E_{23}, they are *dynamically* quite different.

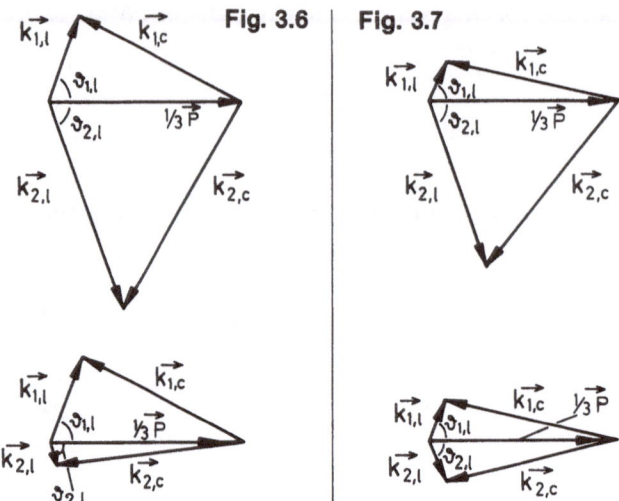

Fig. 3.6. The laboratory momenta of particles 1 and 2 corresponding to the intersections of the right dotted vertical line with the kinematical curve in Fig. 3.5a, b. Shown are also the center-of-mass momenta

Fig. 3.7. The same as in Fig. 3.6 for the left dotted vertical line in Fig. 3.5a, b

3.5.3 Selected Numerical Studies

Let us first regard the multiple scattering series in a simple nuclear model. Low energy, two-nucleon phase shifts can be reproduced fairly well by a separable two-nucleon interaction of the form

$$V(p,p') = \lambda g(p) g(p') \tag{3.317}$$

as was proposed a longtime ago by *Yamaguchi* [3.47]. The so called form factors are simply given by

$$g(p) = \frac{1}{p^2 + \beta^2} . \tag{3.318}$$

For pure *s*-wave interactions there are two channels, $^1S_0$ and $^3S_1$. For each channel there are two parameters available, λ and β, to fit the low energy two-nucleon observables, namely the scattering length and effective range. Since the interactions are attractive, $\lambda < 0$, the resulting theoretical phase shifts stay positive and will therefore necessarily deviate from the experimental ones at higher energies (see Sect. 2.6). This separable force will therefore be too strong. Nevertheless, qualitatively they are useful to explain many features in the three-nucleon system.

Exercise: Use the two-body t matrix for the force (3.317) and relate the two parameters λ and β to scattering length and effective range parameters. Can one reproduce at the same time the deuteron pole and antibound state pole in the states $^3S_1$ and $^1S_0$?

The equations to be solved are the one by Lovelace, (3.283). Before a numerical analysis can be started further steps are to be carried through: symmetrization, partial-wave decomposition and the introduction of a device to treat the "moving singularities" of the transition potential $\langle q_\alpha | Z_{\alpha\beta} | q_\beta \rangle$. Since we shall describe symmetrization and partial-wave decomposition in detail in Sect. 3.7 in the context of the bound state problem, we shall skip it here. The third point is described in the lectures by *Schmidt* and *Ziegelman* [3.48] and in the original articles cited below.

In the partial-wave decomposition and for this simple force, Lovelace's equations reduce to a set of two coupled equations for the state of total spin $S = 1/2$ (doublet) and one equation for the state of total spin $S = 3/2$ (quartet) for each total orbital angular momentum L. The multiple scattering series results by iterating these two sets of integral equations, as we saw in Sect. 3.4.3. Some terms of the Neumann series for one of the two amplitudes in the doublet case, the on-shell, physical, elastic scattering amplitude, are shown in Table 3.1 for $L = 0$, 1 and 2 [3.49]. We have chosen three energies $E^{\text{lab}} =$ 14.4 MeV, 30 MeV and 100 MeV. Whereas the Neumann series converges for $L = 2$ and $L = 1$, the series diverges strongly in the case $L = 0$, and only at $E = 100$ MeV is convergence weakly indicated. This divergence is of course to be expected, since the three-nucleon bound state sits in the state $L = 0$, $S = 1/2$. In Table 3.2 we show the related Padé ratios, which exhibit a nice convergence in all cases. Needless to say, the direct inversion of the one-dimensional integral equation reproduces the numbers found by the Padé method.

In the quartet case, Table 3.3 shows the Neumann series together with the Padé approximants for $L = 0$ and $E = 14.4$ MeV. Table 3.1 clearly exhibits that a low order calculation, like in an impulse approximation, would be meaningless in that model for the energies considered. This is further commented on in [3.40]. Since the divergence is caused by the mere existence of the three-nucleon bound state, the series will also diverge for more realistic two-nucleon forces.

It was *Amado* [3.50] who first derived one-dimensional coupled integral equations of the type (3.283) for a model field theory. Besides *Aaron* and *Amado* [3.51] it was *Phillips* [3.51] who showed that solutions of (3.283) based on simple two-nucleon forces are already capable of reproducing fairly well the experimental three-nucleon data [3.52]. We show in Fig. 3.8 an example of a theoretical break-up spectrum [3.52] in comparison to experimental data. More sophisticated calculations in the framework of separable forces including spin observables have been carried through by *Doleshall* [3.53]. For a review see [3.54].

Table 3.1. Some terms in the Neumann series for the $n-d$ doublet scattering amplitude in the states $L = 0$, 1, and 2 at $E = 14.4$ MeV, 50 MeV and 100 MeV

$E = 14.4$ MeV

n	$L = 0$		$L = 1$	$L = 2$
0	-1.665		0.5956	-0.2070
1	-7.017	$-i\ \ 6.061$	$-0.8840\ -i\ 0.7788$	$-0.1055\ -i\ 0.1055$
2	6.36	$-i\ \ 10.35$	$-0.2458\ +i\ 0.2779$	$0.0100\ -i\ 0.0021$
3	19.88	$-i\ \ \ 1.05$	$0.1353\ -i\ 0.0398$	$0.0006\ -i\ 0.0001$
4	15.91	$+i\ \ 21.61$	$-0.0544\ -i\ 0.0134$	$0.0000\ -i\ 0.0000$
5	-11.71	$+i\ \ 38.90$	$0.0191\ +i\ 0.0178$	
6	-55.43	$+i\ \ 18.83$	$-0.0033\ -i\ 0.0112$	
7	-70.94	$-i\ \ 49.59$	$-0.0013\ +i\ 0.0053$	
8	-2.1	$-i\ 126.4$	$0.0017\ -i\ 0.0017$	
9	148.3	$-i\ 112.0$	$-0.0011\ +i\ 0.0003$	
10	260.5	$+i\ \ 79.3$		
11	130.9	$+i\ 377.6$		
12	-334.9	$+i\ 481.0$		

$E = 50$ MeV

n	$L = 0$		$L = 1$	$L = 2$
0	-0.3594		0.1825	-0.0822
1	-1.682	$-i\ \ 1.405$	$-0.4626\ -i\ 0.4021$	$-0.1141\ -i\ 0.1059$
2	1.814	$-i\ \ 1.327$	$-0.1688\ +i\ 0.0554$	$0.0081\ +i\ 0.0027$
3	1.879	$+i\ \ 1.681$	$0.0495\ +i\ 0.0369$	$0.0008\ +i\ 0.0005$
4	-1.082	$+i\ \ 2.042$	$0.0003\ -i\ 0.0212$	$0.0000\ +i\ 0.0001$
5	-2.327	$-i\ \ 0.479$	$-0.0068\ +i\ 0.0046$	$0.0000\ +i\ 0.0000$
6	-0.156	$-i\ \ 2.314$	$0.0028\ +i\ 0.0010$	
7	2.186	$-i\ \ 0.788$	$-0.003\ \ \ -i\ 0.0011$	
8	1.337	$+i\ \ 1.870$		
9	-1.426	$+i\ \ 1.788$		
10	-2.093	$-i\ \ 0.879$		

$E = 100$ MeV

n	$L = 0$		$L = 1$	$L = 2$
0	-0.1203		0.0714	-0.0368
1	-0.6156	$-i\ \ 0.3840$	$-0.2402\ -i\ 0.1536$	$-0.0836\ -i\ 0.0545$
2	0.5268	$-i\ \ 0.3200$	$-0.0754\ +i\ 0.0133$	$0.0051\ +i\ 0.0018$
3	0.2579	$+i\ \ 0.4589$	$0.0107\ +i\ 0.0202$	$0.0003\ +i\ 0.0005$
4	-0.3141	$+i\ \ 0.1852$	$0.0043\ -i\ 0.0048$	$-0.0000\ +i\ 0.0000$
5	-0.1557	$-i\ \ 0.2331$	$-0.0019\ -i\ 0.0008$	
6	0.1650	$-i\ \ 0.1221$	$-0.0000\ +i\ 0.0006$	
7	0.0981	$+i\ \ 0.1187$	$0.0002\ -i\ 0.0000$	
8	-0.0842	$+i\ \ 0.0771$		
9	-0.0607	$-i\ \ 0.0597$		
10	0.0420	$-i\ \ 0.0474$		

Table 3.2. The Padé summations related to the series of Table 3.1

$E = 14.4\,\text{MeV}$

n	L = 0	L = 1	L = 2
0	− 1.665	0.5956	0.2070
1	− 8.682 − i 6.061	− 0.2883 − i 0.7788	− 0.3124 − i 0.1055
2	− 0.665 − i 6.038	− 0.4324 − i 0.4554	− 0.3027 − i 0.1070
3	− 0.792 − i 1.751	− 0.4387 − i 0.5457	− 0.3018 − i 0.1078
4	3.238 − i 2.230	− 0.4400 − i 0.5446	− 0.3017 − i 0.1078
5	1.006 − i 1.143	− 0.4390 − i 0.5449	− 0.3017 − i 0.1078
6	0.631 − i 1.523	− 0.4378 − i 0.5440	
7	0.598 − i 1.634	− 0.4377 − i 0.5438	
8	0.603 − i 1.645	− 0.4377 − i 0.5438	
9	0.602 − i 1.642	− 0.4377 − i 0.5438	
10	0.603 − i 1.641		
11	0.603 − i 1.641		
12	0.603 − i 1.641		

$E = 50\,\text{MeV}$

n	L = 0	L = 1	L = 2
0	− 0.3594	0.1825	− 0.0822
1	− 2.041 − i 1.405	− 0.2801 − i 0.4021	− 0.1963 − i 0.1059
2	− 0.633 − i 1.346	− 0.4452 − i 0.2888	− 0.1886 − i 0.1032
3	− 0.452 − i 0.863	− 0.4029 − i 0.3269	− 0.1873 − i 0.1026
4	0.002 − i 0.413	− 0.4030 − i 0.3266	− 0.1873 − i 0.1025
5	− 0.279 − i 0.728	− 0.4030 − i 0.3271	− 0.1873 − i 0.1025
6	− 0.293 − i 0.756	− 0.4035 − i 0.3262	
7	− 0.292 − i 0.761	− 0.4035 − i 0.3262	
8	− 0.291 − i 0.761		
9	− 0.292 − i 0.761		
10	− 0.292 − i 0.761		

$E = 100\,\text{MeV}$

n	L = 0	L = 1	L = 2
0	− 0.1203	0.0714	− 0.0368
1	− 0.7359 − i 0.3840	− 0.1688 − i 0.1536	− 0.1203 − i 0.0545
2	− 0.3454 − i 0.4012	− 0.2546 − i 0.1175	− 0.1155 − i 0.0527
3	− 0.2797 − i 0.3166	− 0.2305 − i 0.1253	− 0.1149 − i 0.0521
4	− 0.2579 − i 0.2322	− 0.2308 − i 0.1251	− 0.1150 − i 0.0521
5	− 0.2512 − i 0.2850	− 0.2310 − i 0.1252	− 0.1150 − i 0.0521
6	− 0.2517 − i 0.2884	− 0.2310 − i 0.1251	
7	− 0.2515 − i 0.2887	− 0.2310 − i 0.1251	
8	− 0.2515 − i 0.2887		

Kloet and *Tjon* [3.55] first handled directly local forces without separable approximations by summing the multiple scattering series using the Padé technique. This study for *s*-wave spin-dependent forces was carried through for the elastic and break-up channel.

Separable approximations to local forces of the UPA type, mentioned in Sect. 3.6.1, work very well. This is exemplified in [3.56] and shown in Fig. 3.9.

Table 3.3. Some terms in the Neumann series for the $n-d$ quartet scattering amplitude in the state $L = 0$ at $E = 14.4\,\text{MeV}$ together with its Padé summation

$E = 14.4\,\text{MeV}$		
n	$L = 0$	Padé
0	3.330	3.330
1	$-1.588 - i\,5.956$	$1.742 - i\,5.956$
2	$-6.073 + i\,2.605$	$1.344 - i\,2.112$
3	$4.379 + i\,3.987$	$-0.695 - i\,2.607$
4	$1.267 - i\,5.332$	$-0.856 - i\,2.495$
5	$-4.927 + i\,1.431$	$-0.837 - i\,2.520$
6	$3.457 + i\,3.353$	$-0.838 - i\,2.521$
7	$1.128 - i\,4.381$	$-0.838 - i\,2.521$
8	$-4.100 + i\,1.118$	$-0.838 - i\,2.521$
9	$2.820 + i\,2.826$	$-0.838 - i\,2.521$

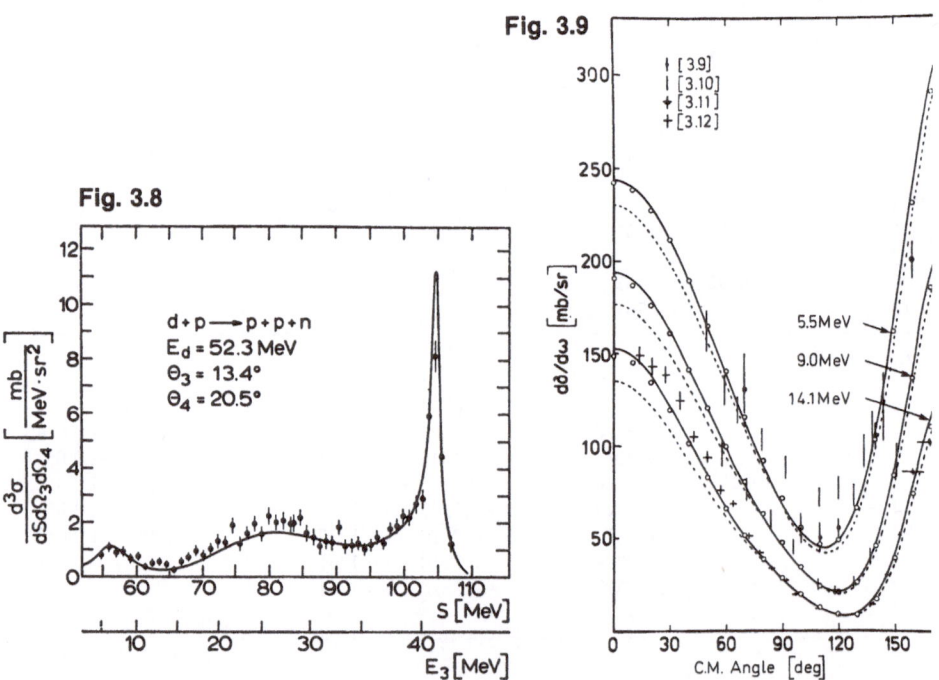

Fig. 3.8

Fig. 3.9

Fig. 3.8. Theoretical and experimental break-up spectrum. The two protons (called 3 and 4) are detected in coincidence. S is the arc length of the kinematic line. The outer peaks are final-state interaction peaks; the middle peak is a spectator peak

Fig. 3.9. Experimental and theoretical differential cross section for $n-d$ scattering at 5.5, 9.0, 14.1 MeV. The full line is the exact result of [3.55]. The dashed line is the result for a Yamaguchi potential. The UPA result is shown by open circles

An interesting study [3.57] explores systematically all kinematical regions in the break-up channel, some of which show high sensitivity to the types of two-nucleon forces used and some very low sensitivity. The last case may be especially interesting in the search for effects of three-nucleon forces.

First calculations with local, two-nucleon forces, which are considered to be more realistic, were done in configuration space [3.20] and in a perturbational scheme in momentum space [3.58]. Certainly one can expect in the near future calculations for scattering processes of the same high standard as for the three-nucleon bound state (see Sect. 3.7). Only then will a conclusive test about the present day description of low energy nuclear dynamics be available. Since the three-nucleon bound state seems to require a three-nucleon force, it may be expected that its effect will also show up in the scattering observables.

One can think of many examples for quasi three-body problems in nuclear physics, some of which have been studied. We mention the system $(d + \alpha)$, where the α-particle is assumed to be elementary. Both ^{6}Li bound states as, well as $(d - \alpha)$ elastic scattering observables (angular distribution, vector and tensor polarisations), were calculated [3.59] with the assumption of separable approximations for the nucleon-nucleon and nucleon-alpha interactions. Though certain features are reproduced noticeable discrepancies are present, for instance in the absolute values of the level spectrum.

3.6 The Three-Nucleon Bound State

We shall illustrate in some detail in this example the techniques required to handle the complexity of a three-nucleon system. On top of the geometrical difficulty of a three-body problem, one faces spin- and isospin degrees of freedom and the violent variation of the nuclear force at short distances which induces high momentum components into the wave function. Various procedures have been developed and are being used: the Ritz-variational treatment [3.60 − 62]; expansion of the wave function into hyperspherical harmonics [3.63] which converts the Schrödinger equation into an infinite set of second order differential equations in one variable, the hyperradius; and the use of the Faddeev equations both in a momentum [3.64 − 66, 3.39, 37] and coordinate space [3.21, 22] representation. We shall present only the momentum space treatment of the Faddeev equations without claiming that the other procedures are or could not be equally powerful enough to handle the problem. The momentum space, however, is the natural one if one uses field theoretical potentials like the OBEP (one-boson-exchange potential) [3.67]. The momentum space representation appears to be unavoidable if one likes to treat relativistic equations [3.68, 38], where expressions like $\sqrt{m^2 + p^2}$ occur.

3.6.1 The Faddeev Equations with a Three-Body Force

In order to describe the general situation we shall include a three-body force, V_4. Then the Schrödinger equation for the bound state reads in integral form

$$|\Psi\rangle = \frac{1}{E - H_0} \left(\sum_{i=1}^{3} V_i + V_4 \right) |\Psi\rangle . \tag{3.319}$$

As we saw in Sect. 3.4.4 it is natural for certain three-nucleon force models to decompose V_4 into three parts

$$V_4 = \sum_{i=1}^{3} V_4^{(i)} . \tag{3.320}$$

Here $V_4^{(1)}$ is supposed to be symmetric under exchange of particles 2 and 3 and $V_4^{(2)}$ and $V_4^{(3)}$ result from cyclical permutations of $V_4^{(1)}$. Therefore the sum in (3.320) will be totally symmetric, an obvious requirement for three identical particles. The decomposition (3.320), inserted into (3.319), suggests grouping together a pair interaction $V_i \equiv V_{jk}$ and the part $V_4^{(i)}$ of V_4, which like V_i is symmetric under exchange of particles j and k. Therefore we write

$$|\Psi\rangle = G_0 \sum_{i=1}^{3} (V_i + V_4^{(i)}) |\Psi\rangle \tag{3.321}$$

and define the Faddeev components as

$$|\psi_i\rangle = G_0 (V_i + V_4^{(i)}) |\Psi\rangle . \tag{3.322}$$

Obviously they sum up to $|\Psi\rangle$:

$$|\Psi\rangle = \sum_{i=1}^{3} |\psi_i\rangle . \tag{3.323}$$

In this manner one avoids the alternative of introducing a fourth Faddeev component linked to V_4. One now inserts the decomposition (3.323) for $|\Psi\rangle$ into the rhs of (3.322) and solves for $|\psi_i\rangle$:

$$|\psi_i\rangle = G_0 T_i \sum_{j \neq i} |\psi_j\rangle . \tag{3.324}$$

The operators T_i obey the integral equations

$$T_i = (V_i + V_4^{(i)}) + (V_i + V_4^{(i)}) G_0 T_i . \tag{3.325}$$

Dropping $V_4^{(i)}$ the operators T_i becomes the two-body operators t_i and (3.324) the usual set of homogeneous, bound state, Faddeev equations.

Because of the presence of $V_4^{(i)}$, Eq. (3.325) are no longer two-body problems. However, according to the present day insight, three-nucleon forces seem to be a correction to two-nucleon forces and the equivalent form

$$T_i = t_i + (1 + t_i G_0) V_4^{(i)} (1 + G_0 T_i) \tag{3.326}$$

is more adequate. It lends itself to a perturbative treatment of $V_4^{(i)}$ which simplifies matters. Thus to first order in $V_4^{(i)}$, one gets

$$T_i \approx t_i + (1 + t_i G_0) V_4^{(i)} (1 + G_0 t_i) . \tag{3.327}$$

Let us now consider that the particles are identical. The three-nucleon state $|\Psi\rangle$ is totally antisymmetric and therefore the components (3.322) have the obvious properties

$$|\psi_2\rangle = P_{12} P_{23} |\psi_1\rangle \tag{3.328}$$

$$|\psi_3\rangle = P_{13} P_{23} |\psi_1\rangle . \tag{3.329}$$

Thus the functional form of only one Faddeev component is needed, say $|\psi_1\rangle$, which obeys

$$|\psi_1\rangle = G_0 T_1 (P_{12} P_{23} + P_{13} P_{23}) |\psi_1\rangle . \tag{3.330}$$

The other two equations are identical. Though we end up with one Faddeev equation, the geometrical complexity of the three-body problem is of course still present in the from of the permutation operators

$$P \equiv P_{12} P_{23} + P_{13} P_{23} . \tag{3.331}$$

Once $|\psi_1\rangle$ is determined, the total state is given by

$$|\Psi\rangle = (1 + P) |\psi_1\rangle . \tag{3.332}$$

3.6.2 Momentum Space Representation

We defined in (3.6b) the states $|p_1 q_1\rangle$, which describe the free relative motions of three particles with the help of the Jacobi momenta (3.3b). An equivalent description is to use the quantum numbers for the relative orbital angular momenta l and λ within the pair (2 3) and between particle 1 and the pair (2 3), respectively, together with the magnitudes of p and q (we drop the index 1). Moreover, we couple l and λ to the total orbital angular momentum L and define the partial-wave states through

$$\langle p'q'|pq(l\lambda)LM\rangle \equiv \frac{\delta(p-q')}{pp'} \frac{\delta(q-q')}{qq'} \mathcal{Y}_{l\lambda}^{LM}(\hat{p}'q') . \tag{3.333}$$

Here $\mathcal{Y}_{l_1 l_2}^{LM}$ is the simultaneous eigenfunction of l_1^2, l_2^2, L^2 and L_z [3.69].

We proceed similarly in the spin space. The spin s of the (2 3) subsystem is coupled with the spin 1/2 of particle 1 to the total spin S accompanied by M_S:

$$|(s\tfrac{1}{2})SM_S\rangle = \sum_{\mu\nu} C(s\tfrac{1}{2}S,\mu\nu M_S)\chi_{s\mu}(2\,3)\chi_{1/2\nu}(1) . \tag{3.334}$$

The state of total isospin quantum numbers TM_T is constructed in complete analogy: $|(t\tfrac{1}{2})TM_T\rangle$. Finally L and S are coupled to the total angular momentum J of the three-nucleon bound state. Thus we get the basis states

$$|pq(l\lambda)L(s\tfrac{1}{2})S(LS)JM(t\tfrac{1}{2})TM_T\rangle$$
$$= \sum_{M_L M_S} C(LSJ,M_L M_S M)|pq(l\lambda)LM_L\rangle |(s\tfrac{1}{2})SM_S\rangle |(t\tfrac{1}{2})TM_T\rangle . \tag{3.335}$$

Instead of working in LS-coupling, one can combine l and s to the total subsystem angular momentum j, and λ and s_1 to the total spectator angular momentum I, which finally combine to J. The two coupling schemes are connected in the following well known manner [3.70]:

$$|pq(ls)j(\lambda\tfrac{1}{2})I(jI)J(t\tfrac{1}{2})T\rangle = \sum_{LS}\sqrt{\hat{j}\hat{I}\hat{L}\hat{S}} \left\{\begin{matrix} l & s & j \\ \lambda & \tfrac{1}{2} & I \\ L & S & J \end{matrix}\right\}$$
$$\times |pq(l\lambda)L(st)S(LS)J(t\tfrac{1}{2})T\rangle . \tag{3.336}$$

We introduced the convenient abbreviation

$$\hat{l} \equiv 2l+1 . \tag{3.337}$$

It remains to guarantee the antisymmetry of $|\Psi\rangle$ as given in (3.332). The Faddeev component $|\psi_1\rangle$, as defined in (3.322) has only a definite symmetry with regard to the pair (2 3). It is antisymmetric under the exchange of particles (2 3). The subsystem states $|lm_l\rangle$, $|sm_s\rangle$, $|tm_t\rangle$ have the well known exchange properties $(-)^l$, $(-)^{s+1}$, and $(-)^{t+1}$, respectively. Therefore the basis states allowed for $|\psi_1\rangle$ are selected through the requirement

$$(-)^{l+s+t} = -1 . \tag{3.338}$$

Once $|\psi_1\rangle$ is chosen antisymmetric in the pair (2 3), the form (3.332) guarantees antisymmetry for $|\Psi\rangle$ in all pairs. Henceforth we shall represent all the discrete quantum numbers in the basis of jI-coupling by α and shall denote the basis states obeying the restriction (3.338) briefly by $|pq\alpha\rangle$.

We face now the problem of representing the single Faddeev equation (3.330) in the basis $|pq\alpha\rangle$. The operator G_0 is diagonal and is simply:

$$\langle pq\alpha|G_0|p'q'\alpha'\rangle = \delta_{\alpha\alpha'}\frac{\delta(p-p')}{pp'}\frac{\delta(q-q')}{qq'}\left(E-\frac{p^2}{m}-\frac{3}{4}\frac{q^2}{m}\right)^{-1}.$$
(3.339)

Thus the first little step yields

$$\langle pq\alpha|\psi_1\rangle = \left(E-\frac{p^2}{m}-\frac{3}{4}\frac{q^2}{m}\right)^{-1}\langle pq\alpha|T_1P\psi_1\rangle.$$
(3.340)

Now T_1 and P are both symmetric in (2 3). Therefore since the state $|pq\alpha\rangle$ in (3.340) is chosen antisymmetric in (2 3), only the antisymmetric states in the completeness relation, when inserted between T_1 and P and between P and $|\psi_1\rangle$, will give a non zero contribution:

$$\langle pq\alpha|\psi_1\rangle = \left(E-\frac{p^2}{m}-\frac{3}{4}\frac{q^2}{m}\right)^{-1}\sum_{\alpha'}\int dp'p'^2\int dq'q'^2$$

$$\times \sum_{\alpha''}\int dp''p''^2\int dq''q''^2\langle pq\alpha|T_1|p'q'\alpha'\rangle$$

$$\times \langle p'q'\alpha'|P|p''q''\alpha''\rangle\langle p''q''\alpha''|\psi_1\rangle.$$
(3.341)

According to (3.326) T_1 has a simple part, the two-body t-operator t_1, whose matrix elements have to be diagonal in the quantum numbers of the spectator particle 1:

$$\langle pq\alpha|t_1(E)|p'q'\alpha'\rangle = \frac{\delta(q-q')}{qq'}\delta_{\lambda\lambda'}\delta_{II'}$$

$$\langle p(ls)jt|\hat{t}_1\left(E-\frac{3}{4}\frac{q^2}{m}\right)|p'(l's)jt\rangle\delta_{jj'}\delta_{tt'}\delta_{ss'}.$$
(3.342)

In addition to the total subsystem angular momentum j, we have also assumed that s and t are conserved. This is a good approximation and is broken only on the level of the electromagnetic interaction.

The second part of T_1 is essentially determined by $V_4^{(1)}$ [see (3.327)] and will be in general a full matrix without zeros [3.37, 42].

We are left with the proper three-body problem which is of a geometrical nature; namely the linking of the three different arrangements of the three particles. Let us indicate explicitly by a subscript to which pair and spectator particle the quantum numbers refer. Thus $|pq\alpha\rangle$ should be denoted by

$$|pq\alpha\rangle \equiv |pq\alpha\rangle_1 \tag{3.343}$$

and we get

$$_1\langle pq\alpha|P_{12}P_{23}+P_{13}P_{23}|p'q'\alpha'\rangle_1 = {}_1\langle pq\alpha|p'q'\alpha'\rangle_2 + {}_1\langle pq\alpha|p'q'\alpha'\rangle_3 . \tag{3.344}$$

Note that all the states in (3.341) are of the type (3.343). The meaning of the overlap matrix elements in (3.344) should be obvious. For instance, the first term is the probability amplitude of finding the relative momenta p and q and the discrete quantum numbers α referring to the arrangement 1, 2 3 in the state in which particles (3 1) and 2 with respect to (3 1) have relative momenta p' and q', respectively, and sit in the discrete states α'.

Evaluating the right hand side of (3.344) we first note that the two overlaps are equal. We use

$$P_{13}P_{23} = P_{23}P_{12}P_{23}P_{23} \tag{3.345}$$

and apply the outer P_{23}'s to the right and to the left in the following manner:

$$\begin{aligned}
_1\langle pq\alpha|p'q'\alpha'\rangle_3 &\equiv {}_1\langle pq\alpha|P_{13}P_{23}|p'q'\alpha'\rangle_1 \\
&= (-)^{l+s+t}(-)^{l'+s'+t'} {}_1\langle pq\alpha|P_{12}P_{23}|p'q'\alpha'\rangle_1 \\
&= {}_1\langle pq\alpha|P_{12}P_{23}|p'q'\alpha'\rangle_1 \equiv {}_1\langle pq\alpha|p'q'\alpha'\rangle_2 .
\end{aligned} \tag{3.346}$$

The phases collapsed to 1 due to the symmetry property (3.338) of our basis states.

The next step is to decouple spin-, isospin- and momentum-space:

$$_1\langle pq\alpha|p'q'\alpha'\rangle_2 = \sum_{LS}\sum_{L'S'}\sqrt{\hat{j}\hat{l}\hat{L}\hat{S}}\sqrt{\hat{j}'\hat{l}'\hat{L}'\hat{S}'}$$

$$\times \begin{Bmatrix} l & s & j \\ \lambda & \frac{1}{2} & I \\ L & S & J \end{Bmatrix} \begin{Bmatrix} l' & s' & j' \\ \lambda' & \frac{1}{2} & I' \\ L' & S' & J \end{Bmatrix} \tag{3.347}$$

$$\times {}_1\langle pq(l\lambda)L|p'q'(l'\lambda')L'\rangle_2 \, {}_1\langle(s\tfrac{1}{2})S|(s'\tfrac{1}{2})S'\rangle_2 \, {}_1\langle(t\tfrac{1}{2})T|(t'\tfrac{1}{2})T\rangle_2 .$$

The spin and isospin matrix elements are recoupling coefficients between three spin 1/2 particles and are easily shown to be [3.70]

$$_1\langle(s\tfrac{1}{2})S|(s'\tfrac{1}{2})S'\rangle_2 = \delta_{SS'}(-)^{s'}\sqrt{\hat{s}\hat{s}'}\begin{Bmatrix} \frac{1}{2} & \frac{1}{2} & s \\ \frac{1}{2} & S & s' \end{Bmatrix} . \tag{3.348}$$

An analogous expression holds for the isospin. The real work lies in the momentum part. Though its calculation is at the very root of the three-body problem, it is tedious and we defer it to the Appendix 3.65. We present here the result:

$$_1\langle pq\alpha|p'q'\alpha'\rangle_2 + {}_1\langle pq\alpha|p'q'\alpha'\rangle_3$$

$$= \int_{-1}^{1} dx \frac{\delta(p-\pi_1)}{p^{l+2}} \frac{\delta(p'-\pi_2)}{p'^{l'+2}} G_{\alpha\alpha'}(qq'x) \tag{3.349}$$

with

$$\pi_1 = \sqrt{q'^2 + \tfrac{1}{4}q^2 + qq'x} \tag{3.350}$$

$$\pi_2 = \sqrt{q^2 + \tfrac{1}{4}q^2 + qq'x} \tag{3.351}$$

and

$$G_{\alpha\alpha'}(qq'x) = \sum_k P_k(x) \sum_{l_1+l_2=l} \sum_{l_1'+l_2'=l'} q^{l_2+l_2'} q'^{l_1+l_1'} g_{\alpha\alpha'}^{kl_1l_2l_1'l_2'}. \tag{3.352}$$

The quantity $g_{\alpha\alpha'}^{kl_1l_2l_1'l_2'}$ is purely geometrical and is given in the Appendix.

We are now in a position to write down the partial-wave representation of the Faddeev equation in momentum space. Using two δ-functions from (3.349) the Faddeev components

$$\psi_\alpha(pq) \equiv \langle pq\alpha|\psi_1\rangle \tag{3.353}$$

obey the infinite set of coupled equations

$$\psi_\alpha(pq) = \left(E - \frac{p^2}{m} - \frac{3}{4}\frac{q^2}{m}\right)^{-1} \sum_{\alpha'\alpha''} \int_0^\infty dq' q'^2 \int_0^\infty dq'' q''^2$$

$$\times \int_{-1}^{1} dx \frac{\langle pq\alpha|T_1|\pi_1 q'\alpha'\rangle}{\pi_1^{l'}} G_{\alpha'\alpha''}(q'q''x) \psi_{\alpha''}\frac{(\pi_2 q'')}{\pi_2^{l''}}. \tag{3.354}$$

The geometrical recoupling reflects itself in the skew arguments π_1 and π_2 in the matrix element of T_1 and the unknown Faddeev components $\psi_{\alpha''}$.

Angular momentum reductions of the Faddeev equations have been given in many papers [3.71].

3.6.3 A Technical Remark

In order to solve the set of integral equations (3.354) one has to approximate the three integrals. This is achieved by choosing appropriate quadrature points in q', x, and q''. With respect to the q-variable one gets immediately a closed set of unknowns. This entails however a very high number of π_2-values according to (3.351), which cannot be handled numerically. Therefore an interpolation in the p-variable seems obligatory. For that purpose, an interpolation algorithm of the form

$$f(x) \approx \sum_i S_i(x) f(x_i) \tag{3.355}$$

appears to be very convenient. Here $S_i(x)$ are known functions and $f(x_i)$ are the (known or unknown) function values at an appropriately choosen set of grid points. We developed such a procedure [3.72] using Spline functions. Thus the unknown Faddeev components under the integral in (3.354) are approximated by

$$\psi_{\alpha''}(\pi_2 q'') \approx \sum_k S_k(\pi_2) \psi_{\alpha''}(p_k q'') . \tag{3.356}$$

Now the skew argument π_2 occurs in known Spline functions. The set of p_k-values has to be chosen sufficiently dense to guarantee the desired quality of interpolation.

Denoting by ω_i the weights according to some quadrature rule (for instance Gauss-Legendre), the approximate discretised representation of (3.354) will be [3.37]

$$\psi_\alpha(p_i q_j) = \left(E - \frac{p_i^2}{m} - \frac{3}{4} \frac{q_j^2}{m} \right)^{-1} \sum_{\alpha''} \sum_k \sum_l$$

$$\times \left(\sum_{\alpha'} \sum_{l'} \omega_l q_l^2 \omega_{l'} q_{l'}^2 \sum_s \omega_s \langle p_i q_j \alpha | T_1 | \pi_1 q_l' \alpha' \rangle \right.$$

$$\left. \times G_{\alpha'\alpha''}(q_{l'} q_l x_s) S_k(\pi_2) \right) \psi_{\alpha''}(p_k q_l) . \tag{3.357}$$

This is a homogeneous algebraic set of equations for the unknowns $\psi_\alpha(p_i q_j)$. Note that the x-integration involves only known functions and can be carried out immediately.

It turns out that the Faddeev component drops relatively fast in the q-variable (typically one needs q in the interval up to $q_{max} \sim 4-6 \text{ fm}^{-1}$). This is not the case in the p-variable, since the two-body t-matrix induces high momentum components. However, according to (3.351) π_2 is bounded by 3/2 q_{max}. Thus in solving (3.357) one has to know ψ_α only in a relatively short interval in p, which helps to keep the number of mesh points small.

The system (3.357) can be solved as described in Sect. 3.4.3.

3.6.4 Physical Remarks About the Triton

Let us come back to physics. Up to now we still face a problem in an infinite number of partial wave states α. If we regard the dominant contribution to T_1, namely the two-body t-matrix as given in (3.342), then the importance of the Faddeev components ψ_α is controlled by the strength of the two-nucleon force in the partial-wave states α. Regarding the two-nucleon phase shifts in Sect. 2.6 we recognize that the NN force is strongest in the states $^1S_0$ and $^3S_1 - {}^3D_1$. As a first approximation one may therefore set the force equal to zero in the remaining states. Because of the definite spin $J = 1/2$ of the triton, this restricts the allowed α-values to a small finite number. We show the allowed quantum numbers in Table 3.4. Depending on whether one keeps only the spectator angular momentum $\lambda = 0$ or $\lambda = 0$ and $\lambda = 2$, one ends up with 3 or 5 coupled integral equations.

It is important to recognize that this small number refers to the Faddeev component and not to the total state $|\Psi\rangle$. According to (3.332), the permutation operator P induces a very large number of states α. In a charge form factor calculation for instance around 40 states are required [3.65] to exhaust $|\Psi\rangle$. As a sideremark we mention that the form factor can be calculated directly from the Faddeev components [3.37]. The small number of partial-wave states is the geometrical advantage of decomposing $|\Psi\rangle$ into Faddeev components.

The inclusion of the three-nucleon force may change how many partial-wave states are important for the Faddeev component. This is presently being studied by various groups for special models of three-nucleon forces. A conclusive answer, however, cannot yet be given at the moment. Calculations in coordinate space [3.73] and in momentum space [3.37] demonstrate the feasibility of solving the Faddeev equations including a three-nucleon force. Though in both cases approximations were involved, they can be expected to be overcome in the very near future.

We show in Table 3.5 the theoretical triton binding energies achieved by different methods for various two-nucleon forces, which fit the two-nucleon data about equally well. We see a nice agreement between different techniques

Table 3.4. The discrete quantum numbers for the dominant partial-wave states of the Faddeev component in the triton

	l	s	j	t	λ	I
1	0	0	0	1	0	1/2
2	0	1	1	0	0	1/2
3	2	1	1	0	0	1/2
4	0	1	1	0	2	3/2
5	2	1	1	0	2	3/2

Table 3.5. Theoretical triton binding energies

NN force	Method	E_{3H} [MeV]
Reid [3.76]	Faddeev equation (momentum space) (5 channels)	−6.98 [3.64] −7.02 [3.65, 39, 37]
	Faddeev equation (coordinate space) (5 channels)	−7.0 [3.21] −7.02 [3.22]
	Faddeev equation (momentum space) (NN forces up to $j \leq 2$)	−7.23 [3.39]
	Variational method	−7.75±0.5 [3.60] −7.3±0.2 [3.61, 62]
OBEP [3.67]	Faddeev equation (momentum space) (5 channels)	−7.5 [3.77]
Paris potential [3.75]	Faddeev equation (momentum space) (5 channels)	−7.30 [3.39]
	Faddeev equation (momentum space) (NN forces up to $j \leq 2$)	−7.38 [3.39]

in the 5 channel calculations. The inclusion of NN forces in higher partial wave states gives a small, though not negligible, contribution of about $\Delta E \approx -200$ keV in the case of the Reid potential.

An important difference between the Reid potential and the one-boson-exchange potentials (OBEP) of [3.67] is that they yield different d-state probabilities for the deuteron, $P_d = 6.47\%$ and $P_d = 5.75\%$, respectively. Clearly the smaller d-state probability is associated with a stronger central force, which is more effective in the more tightly bound triton, and thus produces a larger binding energy [3.74]. The situation is similar for the Paris potential [3.75], which has $P_d = 5.77\%$. In addition, small variations in E_{3H} can be traced back to different fits to the $^1S_0$ phase shift.

The first results based on special models for the three-nucleon force are still too controversial at the moment to be quoted. It is an important test of our understanding of the nuclear dynamics to reproduce the experimental value of $E_{3H} = -8.48$ MeV.

3.6.5 Appendix: The Recoupling Coefficient in Momentum Space

In Sect. 3.6.2 we encountered the matrix element

$$X_{12} \equiv {}_1\langle pq(l\lambda)L | p'q'(l'\lambda')L'\rangle_2$$
$$\equiv {}_1\langle pq(l\lambda)L | P_{12}P_{23} | p'q'(l'\lambda')L'\rangle_1. \tag{A.1}$$

We shall present a straightforward evaluation, which uses only elementary techniques as presented for instance in [3.37]. In the first step we exploit the definition of the partial-wave states (3.333):

$$X_{12} = \int dp_1 \, dq_1 \int dp_1' \, dq_1' \, {}_1\langle pq(l\lambda)LM|p_1q_1\rangle_1 \, {}_1\langle p_1q_1|P_{12}P_{23}|p_1'q_1'\rangle_1$$

$$\times \, {}_1\langle p_1'q_1'|p'q'(l'\lambda')L'M'\rangle_1$$

$$= \int dp_1 \, dq_1 \int dp_1' \, dq_1' \, \mathscr{Y}_{l\lambda}^{LM*}(\hat{p}_1\hat{q}_1) \frac{\delta(p-p_1)}{p^2} \, \frac{\delta(q-q_1)}{q^2}$$

$$\times \, {}_1\langle p_1q_1|P_{12}P_{23}|p_1'q_1'\rangle_1 \, \mathscr{Y}_{l'\lambda'}^{L'M'}(\hat{p}_1'\hat{q}_1') \frac{\delta(p_1'-q')}{p'^2} \, \frac{\delta(q_1'-q')}{q'^2} \, .$$

$$(A.2)$$

The matrix element in (A.2), with respect to the three-momentum states (3.6b), is easily evaluated in the following manner:

$$_1\langle p_1q_1|P_{12}P_{23}|p_1'q_1'\rangle_1 = {}_1\langle p_1q_1|p_1'q_1'\rangle_2$$

$$= {}_1\langle p_1q_1| -\tfrac{1}{2}p_1' + \tfrac{3}{4}q_1', -p_1' - \tfrac{1}{2}q_1'\rangle_1 \, . \qquad (A.3)$$

In the second equality, the state determined by Jacobi momenta of the type 2 is rewritten as a state expressed in Jacobi momenta of the type 1. The linear relation used is of the type (3.5). Then (3.7a) yields

$$_1\langle p_1q_1|p_1'q_1'\rangle_2 = \delta(p_1 + \tfrac{1}{2}p_1' - \tfrac{3}{4}q_1') \, \delta(q_1 + p_1' + \tfrac{1}{2}q_1')$$

$$= \delta(p_1 - \tfrac{1}{2}q_1 - q_1') \, \delta(p_1' + q_1 + \tfrac{1}{2}q_1') \, . \qquad (A.4)$$

That choice of the δ-functions which singles out p_1 and p_1' is obviously only one out of several possibilities, but appears to be the most suitable one in the context of the Faddeev equation (3.330).

With (A.4) we get

$$X_{12} = \int dq_1 \int dq_1' \, \mathscr{Y}_{l\lambda}^{LM*}(\widehat{\tfrac{1}{2}q_1 + q_1'}, \hat{q}_1)$$

$$\times \frac{\delta(p - |\tfrac{1}{2}q_1 + q_1'|)}{p^2} \, \frac{\delta(q - q_1)}{q^2} \, \frac{\delta(p' - |q_1 + \tfrac{1}{2}q_1'|)}{p'^2} \, \frac{\delta(q' - q_1')}{q'^2}$$

$$\times \mathscr{Y}_{l'\lambda'}^{L'M'}(\widehat{-q_1 - \tfrac{1}{2}q_1'}, \hat{q}_1')$$

$$= \int d\hat{q} \int d\hat{q}' \, \mathscr{Y}_{l\lambda}^{LM}(\widehat{\tfrac{1}{2}q + q'}, \hat{q}) \frac{\delta(p - \pi)}{p^2} \, \frac{\delta(p' - \pi')}{p'^2}$$

$$\times \mathscr{Y}_{l'\lambda'}^{L'M'}(\widehat{-q - \tfrac{1}{2}q'}, \hat{q}') \, , \qquad (A.5)$$

where

$$\pi = |\tfrac{1}{2}q + q'| \qquad \text{and}$$

$$\pi' = |q + \tfrac{1}{2}q'| \, . \qquad (A.6)$$

The angular integrations cannot be evaluated directly because of the arguments

$$\widehat{a+b} \equiv \frac{a+b}{|a+b|} \tag{A.7}$$

in the spherical harmonics. It is however an easy exercise to verify the following useful relation:

$$Y_{lm}(a+b) = \sum_{l_1+l_2=l} \frac{a^{l_1} b^{l_2}}{|a+b|^l} \sqrt{\frac{4\pi(2l+1)!}{(2l_1+1)!\,(2l_2+1)!}} \, \mathcal{Y}^{\,lm}_{l_1 l_2}(\hat{a}\hat{b}) \,. \tag{A.8}$$

Exercise: Verify (A.8). *Hint:* Decompose $\exp[ix(a+b)] = \exp(ixa)\exp(ixb)$ into partial waves and regard the limits $|a| \to 0$, $|b| \to 0$.

Equipped with (A.8) we can proceed:

$$\mathcal{Y}^{\,LM}_{l\lambda}(\widehat{\tfrac{1}{2}q+q'},\hat{q}) = \sum_{l_1+l_2=l} \frac{(\tfrac{1}{2}q)^{l_1} q'^{\,l_2}}{|\tfrac{1}{2}q+q'|^l}$$

$$\times \sqrt{\frac{4\pi(2l+1)!}{(2l_1+1)!\,(2l_2+1)!}} \{\mathcal{Y}^{\,l}_{l_1 l_2}(\hat{q}\hat{q}')\, Y_\lambda(\hat{q})\}^{LM}. \tag{A.9}$$

The bracket on the rhs of (A.9) indicates that the angular momenta l and λ are coupled to LM. We recognize that the $\hat{q}$-dependence occurs twice, and can be reduced through another useful relation:

$$\mathcal{Y}^{\,lm}_{l_1 l_2}(\hat{a}\hat{a}) = \sqrt{\frac{\hat{l}_1 \hat{l}_2}{4\pi\hat{l}}} \, C(l_1 l_2 l, 00) \, Y_{lm}(\hat{a}) \,, \tag{A.10}$$

where $\hat{l} \equiv 2l+1$.

Exercise: Verify (A.10).

In order to use (A.10) we have to bring together the $\hat{q}$-dependence in (A.9), which amounts to a recoupling of the three angular momenta l_1, l_2, and λ. This is achieved through $6j$-symbols:

$$\{\mathcal{Y}^{\,l}_{l_1 l_2}(\hat{q}\hat{q}')\, Y_\lambda(\hat{q})\}^{LM}$$

$$= \sum_f (-)^{l+\lambda-L}\sqrt{\hat{l}\hat{f}} \begin{Bmatrix} l_2 & l_1 & l \\ \lambda & L & f \end{Bmatrix} \{Y_{l_2}(\hat{q}')\, \mathcal{Y}^{\,f}_{l_1\lambda}(\hat{q}\hat{q})\}^{LM}$$

$$= \sum_f (-)^{l+\lambda-L}\sqrt{\hat{l}\hat{l}_1\hat{\lambda}}\,\frac{1}{\sqrt{4\pi}} \begin{Bmatrix} l_2 & l_1 & l \\ \lambda & L & f \end{Bmatrix} C(l_1\lambda f,00)\, \mathcal{Y}^{\,LM}_{l_2 f}(\hat{q}'\hat{q}) \,. \tag{A.11}$$

In the last step we used (A.10). Thus altogether, (A.9) can be rewritten as

$$\mathscr{Y}_{l\lambda}^{LM}(\widehat{\tfrac{1}{2}q+q'},\hat{q})$$

$$= \sum_{l_1+l_2=l} \frac{(\tfrac{1}{2}q)^{l_1}q'^{l_2}}{|\tfrac{1}{2}q+q'|^l}\sqrt{\frac{(2l+1)!}{(2l_1+1)!\,(2l_2+1)!}}\,(-)^{l+\lambda-L}\sqrt{\hat{l}\,\hat{l}_1\,\hat{\lambda}}$$

$$\times \sum_f \left\{\begin{array}{ccc} l_2 & l_1 & l \\ \lambda & L & f \end{array}\right\} C(l,\lambda f,00)\,\mathscr{Y}_{l_2 f}^{LM}(\hat{q}'\hat{q})\,.\tag{A.12}$$

In exactly the same manner we find

$$\mathscr{Y}_{l'\lambda'}^{L'M'}(\widehat{-q-\tfrac{1}{2}q'},\hat{q}')$$

$$= (-)^{l'}\sum_{l_1'+l_2'=l'} \frac{(\tfrac{1}{2}q)^{l_1'}q'^{l_2'}}{|q+\tfrac{1}{2}q'|^{l'}}\sqrt{\frac{(2l'+1)!}{(2l_1'+1)!\,(2l_2'+1)!}}\,\sqrt{\hat{l}'\,\hat{l}_1'\,\hat{\lambda}'}$$

$$\times \sum_{f'} \left\{\begin{array}{ccc} l_2' & l_1' & l' \\ \lambda' & L' & f' \end{array}\right\} C(l_1'\lambda'f',00)\,\mathscr{Y}_{l_2' f'}^{L'M'}(\hat{q}'\hat{q})\,.\tag{A.13}$$

The remaining angular dependence is due to the dependence on π and π':

$$\frac{\delta(p-\pi)}{p^{l+2}}\,\frac{\delta(p'-\pi')}{p'^{l'+2}} = \sum_k 2\pi\sqrt{\hat{k}}(-)^k g_k\,\mathscr{Y}_{kk}^{00}(\hat{q}'\hat{q})\tag{A.14}$$

with

$$g_k \equiv \int_{-1}^{1} dx P_k(x)\,\frac{\delta(p-\sqrt{\tfrac{1}{4}q^2+q'^2+qq'x})}{p^{l+2}}\,\frac{\delta(p'-\sqrt{\tfrac{1}{4}q'^2+q^2+qq'x})}{p'^{l'+2}}\,.\tag{A.15}$$

We are left with triple products of spherical harmonics. A convenient step is therefore to reduce first a product of two spherical harmonics to one with the aid of (A.10). We have

$$\mathscr{Y}_{kk}^{00}(\hat{q}'\hat{q})\,\mathscr{Y}_{f'l_2'}^{L'M'}(\hat{q}'\hat{q})$$

$$= \sum_{f_1 f_2}\sqrt{\hat{L}'\hat{f}_1\hat{f}_2}\left\{\begin{array}{ccc} k & k & 0 \\ f' & l_2' & L' \\ f_1 & f_2 & L' \end{array}\right\}\{\mathscr{Y}_{kf_1}(\hat{q}'\hat{q})\,\mathscr{Y}_{kl_2'}(\hat{q}\hat{q})\}^{L'M'}$$

$$= \sum_{f_1 f_2}\sqrt{\hat{L}'\hat{f}_1\hat{f}_2}\,\frac{(-)^{f_1+l_2'+k+L'}}{\sqrt{\hat{k}\hat{L}'}}\left\{\begin{array}{ccc} f_2 & f_1 & L' \\ f' & l_2' & k \end{array}\right\}$$

$$\times \sqrt{\frac{\hat{k}\hat{f}'}{4\pi\hat{f}_1}}C(kf'f_1,00)\sqrt{\frac{\hat{k}\hat{l}_2'}{4\pi\hat{f}_2}}C(kl_2'f_2,00)\,\mathscr{Y}_{f_1 f_2}^{L'M'}(\hat{q}'\hat{q})$$

$$= \frac{1}{4\pi}\sqrt{k\hat{f}'\hat{l}_2'}\,(-)^{f'+l_2'+L'}\sum_{f_1 f_2}\begin{Bmatrix} f_2 & f_1 & L' \\ f' & l_2' & k \end{Bmatrix}$$

$$\times C(kf'f_1,00)\,C(kl_2'f_2,00)\,\mathcal{Y}_{f_1 f_2}^{L'M'}(\hat{q}'\hat{q}).\tag{A.16}$$

Now we can do the angular integrations:

$$\int d\hat{q}\,d\hat{q}'\;\mathcal{Y}_{l_2 f}^{LM*}(\hat{q}'\hat{q})\,\mathcal{Y}_{kk}^{00}(\hat{q}'\hat{q})\,\mathcal{Y}_{f'l_2'}^{L'M'}(\hat{q}'\hat{q})$$

$$= \frac{1}{4\pi}\sqrt{k\hat{f}'\hat{l}_2'}\,(-)^{f'+l_2'+L}\begin{Bmatrix} f & l_2 & L \\ f' & l_2' & k \end{Bmatrix} C(kf'l_2,00)\,C(kl_2'f,00)\,\delta_{LL'}\delta_{MM'}\,.\tag{A.17}$$

Collecting finally all the intermediate steps we end up with

$${}_1\langle pq(l\lambda)LM|p'q'(l'\lambda')L'M'\rangle_2$$

$$= \delta_{LL'}\delta_{MM'}\sum_k g_k \sum_{l_1+l_2=l}\;\sum_{l_1'+l_2'=l'} q^{l_2+l_2'}q'^{l_1+l_1'}(-)^{l'}\sqrt{\hat{l}\,\hat{\lambda}\hat{l}'\,\hat{\lambda}'}\,\hat{k}(\tfrac{1}{2})^{l_2+l_1'+1}$$

$$\times \sqrt{\frac{(2l+1)!}{(2l_1)!\,(2l_2)!}}\sqrt{\frac{(2l'+1)!}{(2l_1')!\,(2l_2')!}}\sum_{ff'}\begin{Bmatrix} l_1 & l_2 & l \\ \lambda & L & f \end{Bmatrix} C(l_2\lambda f,00)$$

$$\times \begin{Bmatrix} l_2' & l_1' & l' \\ \lambda' & L & f' \end{Bmatrix} C(l_1'\lambda'f',00)\begin{Bmatrix} f & l_1 & L \\ f' & l_2' & k \end{Bmatrix} C(kl_1f',00)\,C(kl_2'f,00)\,,\tag{A.18}$$

where g_k is given in (A.15).

The forms (3.349) and (3.352) arise if we include the recoupling (3.347) and the spin- and isospin matrix elements of the type (3.348).

The final result is:

$$g_{\alpha\alpha'}^{kl_1 l_2 l_1' l_2'} = \sqrt{\hat{l}\hat{s}\hat{j}\hat{t}\,\hat{\lambda}\hat{l}\hat{l}'\,s'\hat{j}'\hat{t}'\,\hat{\lambda}'\hat{l}'}\,(-)\begin{Bmatrix} \tfrac{1}{2} & \tfrac{1}{2} & t \\ \tfrac{1}{2} & T & t' \end{Bmatrix}$$

$$\times \sum_{LS}\hat{L}\hat{S}\begin{Bmatrix} \tfrac{1}{2} & \tfrac{1}{2} & s \\ \tfrac{1}{2} & S & s' \end{Bmatrix}\begin{Bmatrix} l & s & j \\ \lambda & \tfrac{1}{2} & I \\ L & S & J \end{Bmatrix}\begin{Bmatrix} l' & s' & j' \\ \lambda' & \tfrac{1}{2} & I' \\ L & S & J \end{Bmatrix}\hat{k}(\tfrac{1}{2})^{l_2+l_1'}$$

$$\times \sqrt{\frac{(2l+1)!}{(2l_1)!\,(2l_2)!}}\sqrt{\frac{(2l'+1)!}{(2l_1')!\,(2l_2')!}}\sum_{ff'}\begin{Bmatrix} l_1 & l_2 & l \\ \lambda & L & f \end{Bmatrix} C(l_2\lambda f,00)$$

$$\times \begin{Bmatrix} l_2' & l_1' & l' \\ \lambda' & L & f' \end{Bmatrix} C(l_1'\lambda'f',00)\begin{Bmatrix} f & l_1 & L \\ f' & l_2' & k \end{Bmatrix} C(kl_1f',00)\,C(kl_2'f,00)\,.\tag{A.19}$$

4. Four Interacting Particles

The "generalization" of the Faddeev equations to four and more particles is not trivial. It was *Yakubowsky* [4.1] who discovered a way to set up a system of coupled integral equations, which is in unique correspondence to the Schrödinger equation and has a kernel which gets connected after a certain finite number of iterations. An equivalent coupling scheme was given originally [4.2] by *Grassberger* and *Sandhas*, who applied the quasi-particle method during the derivation and later by *Alt* et al. [4.3], who performed first the operator algebra and afterwards the pole approximations underlying the quasi particle method. This chapter serves to introduce some of these basic ideas, which are necessary to formulate the *N*-body version and which are already present for four particles.

4.1 The Fundamental Set of Lippmann-Schwinger Equations

Whereas for three particles the number of pair interactions, the number of two-fragment channels, the number of Lippmann-Schwinger equations necessary and sufficient for a unique definition of the scattering states, the number of Faddeev equations etc. are all three, corresponding simple coincidences for four (and more particles) no longer hold. There are *six* pair interactions. Then there are *seven* two-fragment channels, which occur now in two types i, (jkl) and (ij), (kl). We show them explicitly:

$$
\begin{array}{l}
1,2\ 3\ 4 \quad
\begin{cases}
1,2,3\ 4 \\
1,3,2\ 4 \\
1,4,2\ 3
\end{cases} \\[2em]
2,3\ 4\ 1 \quad
\begin{cases}
2,3,4\ 1 \\
2,4,3\ 1 \\
2,1,3\ 4
\end{cases} \\[2em]
3,4\ 1\ 2 \quad
\begin{cases}
3,4,1\ 2 \\
3,1,4\ 2 \\
3,2,4\ 1
\end{cases}
\end{array}
\tag{4.1}
$$

$$
\begin{array}{lll}
4,1\,2\,3 & \Longleftarrow &
\begin{array}{l}
4,1,2\ 3 \\
4,2,1\ 3 \\
4,3,1\ 2
\end{array}
\\[2em]
1\ 2,3\ 4 & \Longleftarrow &
\begin{array}{l}
1,2,3\ 4 \\
1\ 2,3,4
\end{array}
\\[1.5em]
1\ 3,2\ 4 & \Longleftarrow &
\begin{array}{l}
1,3,2\ 4 \\
1\ 3,2,4
\end{array}
\\[1.5em]
1\ 4,2\ 3 & \Longleftarrow &
\begin{array}{l}
1,4,2\ 3 \\
1\ 4,2,3\ .
\end{array}
\end{array}
\tag{4.1}
$$

The groupings into two clusters can be further split into three-cluster fragmentations, as worked out above. We recognize that each case of three-fragments occurs three times. For instance 1,2,3 4 can be gained out of 1,2 3 4; 2,3 4 1; and 1 2,3 4. In such a case we say that the three-fragment channel is contained in the two-fragment channel. Since three fragments are characterized by a pair, there are six different channels of that type. Finally one has of course one four-body channel: 1,2,3,4. Experimentally the two-body fragmentation channels are singled out as entrance channels.

It is convenient to introduce the following notation. Two- and three-fragment channels will be denoted by $a_2, b_2, \dots$ and $a_3, b_3, \dots$, respectively. Then $a_3 \subset a_2$ means that the pair characterized by a_3 occurs within a cluster of a_2. (In the above example $a_3 = 3\ 4$ and $a_2 = 1,2\ 3\ 4$; $2,3\ 4\ 1$; $1\ 2,3\ 4$). Correspondingly $a_3 \not\subset a_2$ denotes a situation like $a_2 = 1\ 2$, $a_2 = 1\ 3\ 4,2$.

As for three particles we again introduce channel hamiltonians. Among them, the ones for two-body fragmentations are especially important:

$$
H_{a_2} = H_0 + V_{a_2}
\tag{4.2}
$$

with

$$
V_{a_2} = \sum_{a_2 \subset a_2} V_{a_3} .
\tag{4.3}
$$

For example

$$
V_{1,234} = V_{23} + V_{34} + V_{24}
$$
$$
V_{12,34} = V_{12} + V_{34} .
\tag{4.4}
$$

There are now two types of two-fragment channel states

$$
|\Phi_{a_2}\rangle =
\begin{cases}
|\chi_{a_2}\rangle |q_{a_2}\rangle \\
|\varphi_{a_3}\rangle |\varphi_{b_3}\rangle |q'_{a_2}\rangle .
\end{cases}
\tag{4.5}
$$

Here $|\chi_{a_2}\rangle$ is a three-body bound state and $|q_{a_2}\rangle$ the momentum eigenstate of relative motion between the two fragments.

In the second case $|\varphi_{a_3}\rangle$ and $|\varphi_{b_3}\rangle$ are two-body bound states of the pairs a_3 and b_3 which make up a_2, and $|q'_{a_2}\rangle$ is again the momentum eigenstate of relative motion of the two fragments. Clearly we have

$$H_{a_2}|\phi_{a_2}\rangle = E|\phi_{a_2}\rangle \tag{4.6}$$

with

$$E = E_{a_2} + \mathrm{const}\, q_{a_2}^2$$
$$= E_{a_3} + E_{b_3} + \mathrm{const}'\, q_{a_2}'^{\,2}, \tag{4.7}$$

where E_{a_2}, E_{a_3} are bound state energies and the constants depend on the normalization of the Jacobi momenta.

Finally the intercluster interaction or channel interaction, V^{a_2}, completes the total hamiltonian

$$H = H_{a_2} + V^{a_2}. \tag{4.8}$$

In the above examples

$$V^{1,234} = V_{12} + V_{13} + V_{14}$$
$$V^{12,34} = V_{13} + V_{14} + V_{23} + V_{24}. \tag{4.9}$$

Each of the $7 + 6 + 1 = 14$ channels can be the initial channel for a scattering process, as well as being the final channel if the energy is sufficiently high. Therefore, there will be 14 different types of scattering states.

Let us first focus on the seven states initiated by the various two-body fragmentations. They are defined by

$$|\Psi_{a_2}^{(+)}\rangle = \lim_{\varepsilon \to 0} i\varepsilon\, G(E + i\varepsilon)|\phi_{a_2}\rangle \tag{4.10}$$

with $G(z)$ being the full resolvent operator. Inserting the resolvent identity between G and the channel resolvent operator

$$G_{b_2}(z) \equiv \frac{1}{z - H_{b_2}} \tag{4.11}$$

namely

$$G = G_{b_2} + G_{b_2} V^{b_2} G \tag{4.12}$$

we get

$$|\Psi_{a_2}^{(+)}\rangle = \lim_{\varepsilon \to 0} i\varepsilon\, G_{b_2}(E + i\varepsilon)|\phi_{a_2}\rangle + G_{b_2}V^{b_2}|\Psi_{a_2}^{(+)}\rangle \,. \tag{4.13}$$

In very much the same manner as in Sect. 3.2 one may show that

$$\lim_{\varepsilon \to 0} i\varepsilon\, G_{b_2}|\phi_{a_2}\rangle = \delta_{a_2 b_2}|\phi_{a_2}\rangle \,. \tag{4.14}$$

Exercise: Verify (4.14).

A mathematically rigorous study of (4.14) can be found in Ref. [4.4], which also contains references to previous studies of this problem.

Therefore the scattering states $|\Psi_{a_2}^{(+)}\rangle$ obey the set of seven Lippmann-Schwinger equations out of which six are homogeneous:

$$|\Psi_{a_2}^{(+)}\rangle = |\phi_{a_2}\rangle\, \delta_{a_2 b_2} + G_{b_2}V^{b_2}|\Psi_{a_2}^{(+)}\rangle \,. \tag{4.15}$$

This set is again necessary and sufficient for a unique definition of $|\Psi_{a_2}^{(+)}\rangle$. The proof follows the one for three particles [4.5]. First every solution of (4.15) has to be a solution of the Schrödinger equation. Obviously we can exclude the six states $|\Psi_{c_2}^{(+)}\rangle$, $c_2 \neq a_2$, since they would require a different driving term. As remaining candidates we are left with the seven scattering states which are initiated through three- or four-fragmentation channels. The four-body bound states are of course excluded on energy considerations. Let

$$|\phi_{a_3}\rangle = |\varphi_{a_2}\rangle|q_{a_3}^{(1)}\rangle|q_{a_3}^{(2)}\rangle = |\varphi_{a_3}\rangle|q_{b_3}^{(1)}\rangle|q_{b_3}^{(2)}\rangle \tag{4.16}$$

be a three-fragment channel state. The momentum states describe the two relative motions, which can be either one of the two types shown in Fig. 4.1. The resulting four-body scattering state is

$$|\Psi_{a_3}^{(+)}\rangle = \lim_{\varepsilon \to 0} i\varepsilon\, G(E + i\varepsilon)|\phi_{a_3}\rangle \,. \tag{4.17}$$

Using again (4.12), we have to evaluate $i\varepsilon\, G_{b_2}|\phi_{a_3}\rangle$ in the limit $\varepsilon \to 0$. We can distinguish between $a_3 \subset b_2$ and $a_3 \not\subset b_2$ and find easily as a first step

$$\lim_{\varepsilon \to 0} i\varepsilon\, G_{b_2}|\phi_{a_3}\rangle = 0 \quad \text{for} \quad a_3 \not\subset b_2 \,. \tag{4.18}$$

Exercise: Verify (4.18).

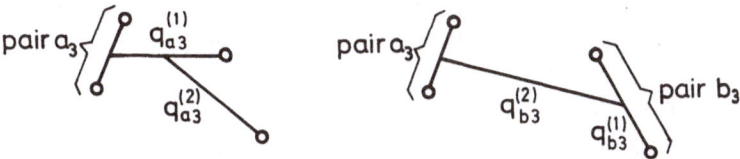

Fig. 4.1. The two types of natural choices of Jacobi variables in the four-body system

The second step is on known ground. For $b_2 = i$, (jkl) and $a_3 \subset b_2$ we find

$$\lim_{\varepsilon \to 0} i\varepsilon\, G_{b_2} |\varphi_{a_3}\rangle\, |q_{a_3}^{(1)}\rangle\, |q_{a_3}^{(2)}\rangle = |\chi_{a_2,q_{a_3}^{(1)}}^{(+)}\rangle\, |q_{a_3}^{(2)}\rangle \equiv |\phi_{b_2,a_3}^{(+)}\rangle \,, \qquad (4.19)$$

where $|\chi_{a_3,q_{a_3}^{(1)}}^{(+)}\rangle$ is a three-body scattering state for the particles (jkl) at the energy E-const $(q_{a_1}^{(2)})^2$. This is a simple consequence of the fact that for H_{b_2}, the fourth particle, described by its relative motion $q_{a_3}^{(2)}$, is a spectator. Once this is accounted for in G_{b_2}, one is left with a three-body scattering problem initiated through the channel state $|\varphi_{a_3}\rangle\, |q_{a_3}^{(1)}\rangle$. In case $b_2 = (ij)(kl)$ and $a_3 \subset b_2$, we find

$$\lim_{\varepsilon \to 0} i\varepsilon\, G_{b_2} |\varphi_{a_3}\rangle\, |q_{b_3}^{(1)}\rangle\, |q_{b_3}^{(2)}\rangle = |\varphi_{a_3}\rangle\, |\phi_{b_3}^{(+)}\rangle\, |q_{b_3}^{(2)}\rangle \equiv |\phi_{b_2,a_3}^{(+)}\rangle \,. \qquad (4.20)$$

Here we used the fact that H_{b_2} is a sum of two pair Hamiltonians ($a_3, b_3 \subset b_2$) and the free one of relative motion of the pairs linked to the momentum $q_{b_3}^{(2)}$. Therefore once the eigenvalues E_{a_3} and const $(q_{b_3}^{(2)})^2$ are inserted into the denominator of G_{b_2}, one is left with a two-body scattering problem, which yields $|\phi_{b_3}^{(+)}\rangle$.

Altogether we come to the conclusion, that the four-body scattering states initiated by the six three-body fragmentation channels obey three inhomogeneous and four homogeneous equations

$$|\Psi_{a_3}^{(+)}\rangle = |\phi_{b_2 a_3}^{(+)}\rangle\, \delta(a_3 \subset b_2) + G_{b_2} V^{b_2} |\Psi_{a_3}^{(+)}\rangle \,. \qquad (4.21)$$

The driving terms are given in (4.19, 20) and the Kronecker-type symbol is due to (4.18). These six states therefore do not satisfy the set (4.15).

It remains to consider the scattering states

$$|\Psi_0^{(+)}\rangle = \lim_{\varepsilon \to 0} i\varepsilon\, G(E + i\varepsilon) |\phi_0\rangle \,, \qquad (4.22)$$

where

$$|\phi_0\rangle = |p_{a_3}\rangle\, |q_{a_3}^{(1)}\rangle\, |q_{a_3}^{(2)}\rangle \equiv |p_{a_3}\rangle\, |q_{b_3}^{(1)}\rangle\, |q_{b_3}^{(2)}\rangle \qquad (4.23)$$

describes four free particles, and the relative motions are described in the two manners shown in Fig. 4.1. Obviously in this case we find

$$\lim_{\varepsilon \to 0} i\varepsilon\, G_{b_2} |\phi_0\rangle \equiv |\phi_0^{(+)}\rangle$$

$$= \begin{cases} |\chi_0^{(+)}\rangle |q_{a_3}^{(2)}\rangle & \text{for} \quad b_2 = i(jkl) \\ |\varphi_{p_3}^{(+)}\rangle |\varphi_{q_{b_3}^{(1)}}^{(+)}\rangle |q_{b_3}^{(2)}\rangle & \text{for} \quad b_2 = (ij)(kl) . \end{cases} \tag{4.24}$$

Here $|\chi_0^{(+)}\rangle$ is a three-particle scattering state initiated by three free particles, whereas the $|\varphi^{(+)}\rangle$ are two-body scattering states. As a consequence, $|\Psi_0^{(+)}\rangle$ always obeys inhomogeneous equations:

$$|\Psi_0^{(+)}\rangle = |\phi_0^{(+)}\rangle + G_{b_2} V^{b_2} |\Psi_0^{(+)}\rangle . \tag{4.25}$$

A very transparent discussion of various Lippmann-Schwinger equations for four particles has been given by *Sandhas* [4.6].

4.2 Coupled Equations in Dummy Variables

In Sect. 3.3.3 we considered three different partial wave decompositions of the three-body wave function, which were directly linked to the three choices of Jacobi coordinates and which guided us to turn the three uncoupled Lippmann-Schwinger equations (3.51 a – c) into the coupled set (3.133). A further decisive step was to then introduce dummy variables and write it in the form (3.135).

Though it formally appears as a system for three unknown functions, that set automatically guarantees that the new indices are only dummy variables and the unique solution is the physical one: $|\Psi^{(1)}\rangle = |\Psi^{(2)}\rangle = |\Psi^{(3)}\rangle = |\Psi\rangle$. We reproduce that system (3.135) in matrix notation

$$\Psi_\alpha = \phi_\alpha + \mathcal{G}_0 \mathcal{V} \Psi_\alpha . \tag{4.26}$$

The new quantities are

$$\mathcal{G}_{0,\mu\nu} = \delta_{\mu\nu} G_\mu \tag{4.27}$$

$$\mathcal{V}_{\mu\nu} = \bar{\delta}_{\mu\nu} V_\nu \tag{4.28}$$

$$\Psi_\alpha^T = (|\Psi_\alpha^{(1)}\rangle, |\Psi_\alpha^{(2)}\rangle, |\Psi_\alpha^{(3)}\rangle) \tag{4.29}$$

and

$$|\phi_\alpha^\mu\rangle = \delta_{\mu\alpha} |\phi_\alpha\rangle . \tag{4.30}$$

Let us now regard the system of seven Lippmann-Schwinger equations (4.15), based on the seven two-fragmentation channels (4.1). The natural Jacobi variables for a fragmentation of the type $i(jkl)$ are the three different choices of Jacobi variables in the three-body cluster (ijk), supplemented by the relative coordinate of particle i with respect to the center-of-mass of the three-body cluster. This is shown in Fig. 4.1 and corresponds to the splittings worked out in (4.1). This yields $4 \times 3 = 12$ different sets of Jacobi variables. For a two-body fragmentation of the type $(ij)(kl)$, the natural variables are different ones as shown in Fig. 4.1. In the same spirit as in Sect. 3.3.3, we may think of a corresponding number of different types of partial-wave states linked to the different types of Jacobi variables.

Now the rhs of (4.15) are

$$G_{b_2} \sum_{c_3 \not\subset b_2} V_{c_3} |\Psi\rangle .$$

In contrast to the three-body system, the pair c_3 does not suggest a unique choice of Jacobi variables. As seen from (4.1), each pair c_3 can occur in three different clusters of two-body fragmentation, and therefore the product $V_{c_3} |\Psi\rangle$ alone would not give a unique handle for chosing the set of Jacobi variables in $|\Psi\rangle$. But the kernel contains additional information through G_{b_2}. Clearly the eigenstates to H_{b_2} required to represent G_{b_2}, will be expressed in the natural variables of b_2, which are not unique but are distinguished by choosing a certain pair $b_3 \subset b_2$. This now provides an argument for selecting the set of Jacobi variables in $|\Psi\rangle$ for $V_{c_3} |\Psi\rangle$. The pair b_3, together with the pair c_3, determines uniquely a two-fragmentation channel c_2 such that $b_3 \subset c_2$ and $c_3 \subset c_2$. Certainly from a practical point of view, overlaps in $G_{b_2} V_{c_3} |\Psi\rangle$ will be most easily calculated in such a manner. Guided by this pragmatic point of view, one chooses a $b_3 \subset b_2$, which together with c_3, fixes $c_2 (\supset b_3, c_3)$ and introduces dummy variables in the following manner

$$G_{b_2} V_{c_3} |\Psi\rangle \equiv G_{b_2} V_{c_3} |\Psi^{c_2 c_3}\rangle . \tag{4.31}$$

Thus we get for instance

$$G_{1,234}(V_{12} + V_{13} + V_{14}) |\Psi\rangle$$
$$\equiv G_{1,234}^{(2,3)}(V_{12} |\Psi^{123,4;12}\rangle + V_{13} |\Psi^{123,4;13}\rangle + V_{14} |\Psi^{14,23;14}\rangle) . \tag{4.32}$$

In this example the pair $b_3 = (2\,3)$ has been selected out of b_2. The other two choices would have led to different chains of dummy indices for $|\Psi\rangle$. What will be the natural choice of dummy variables for $|\Psi\rangle$ on the left hand side? This is again obvious from a practical point of view. One will choose that set of Jacobi coordinates, which is used in the resolvent operator G_{b_2}. Thus the example above would determine $|\Psi^{1,234;23}\rangle$. Finally we recognize that on the right hand side of (4.32), the unknowns $|\Psi^{c_2 c_3}\rangle$ for $c_2 = 1\,2\,3,4$ occur for two choices of $c_3 \subset c_2$. Thus all the different choices of pairs b_3 selected out of b_2 are really necessary to get a closed set in the unknowns. How many unknowns do we have? Obviously $4 \times 3 + 3 \times 2 = 18$, which counts the number of possibilities of selecting an internal pair in the two-fragment clusters. Thus we end up with 18 coupled equations.

The coupling scheme introduced above, and expressed in the example (4.32), can be rephrased in the following equality

$$\sum_{c_3 \nsubseteq b_2} V_{c_3} = \sum_{c_2 \supset b_3} \bar{\delta}_{c_2 b_2} \sum_{c_3 \subset c_2} \bar{\delta}_{c_3 b_3} V_{c_3} \qquad \text{for all } b_3 \subset b_2 . \tag{4.33}$$

This is a specific example of distribution properties of generalized residual interactions [4.7]. We may now insert (4.33) into the set of Lippmann-Schwinger equations (4.15) to get

$$|\Psi_{a_2}^{(+)}\rangle = |\phi_{a_2}\rangle \delta_{a_2 b_2} + G_{b_2} \sum_{\substack{c_2 \supset b_2 \\ (b_3 \subset b_2)}} \bar{\delta}_{b_2 c_2} \sum_{c_3 \subset c_2} \bar{\delta}_{c_3 b_3} V_{c_3} |\Psi_{a_2}^{(+)}\rangle . \tag{4.34}$$

Then we introduce the dummy variables as described above

$$|\Psi_{a_2}^{b_2 b_3}\rangle = |\phi_{a_2}^{b_2 b_3}\rangle + G_{b_2} \sum_{\substack{c_2 \supset b_3 \\ (b_3 \subset b_2)}} \bar{\delta}_{b_2 c_2} \sum_{c_3 \subset c_2} \bar{\delta}_{c_3 b_3} V_{c_3} |\Psi_{a_2}^{c_2 c_3}\rangle . \tag{4.35}$$

The driving term is defined as prescribed in (4.34):

$$|\phi_{a_2}^{b_2 b_3}\rangle = \delta_{a_2 b_2} |\phi_{a_2}\rangle . \tag{4.36}$$
$$\phantom{|\phi_{a_2}^{b_2 b_3}\rangle =} {\scriptstyle (b_3 \subset b_2)}$$

This is a set of eighteen coupled equations which is equivalent to the original one (4.15) if the only solution is $|\Psi_{a_2}^{b_2 b_3}\rangle \equiv |\Psi_{a_2}^{(+)}\rangle$, independent of the artificially introduced cluster indices. We shall give an argument below which strongly weakens this hope. Of course the physical state *is* a solution of this set.

It was *Sandhas* [4.8] who showed how to rewrite the system (4.35) as one which has this desirable property. In (4.27, 28) we introduced the matrices $\mathcal{G}_0$ and $\mathcal{V}$ referring to a three-body subsystem, say b_2. Like the resolvent identity in a two-body system involving G, G_0 and V, one may introduce a resolvent identity

$$\mathcal{G} = \mathcal{G}_0 + \mathcal{G}_0 \mathcal{V} \mathcal{G} , \tag{4.37}$$

which defines a matrix operator $\mathcal{G}$. It follows from this equation that

$$\sum_{c_3 \subset b_2} \mathcal{G}_{b_3 c_3} = G_{b_2} . \tag{4.38}$$

The proof is simple. In explicit notation (4.37) leads to

$$\sum_{c_3 \subset b_2} \mathscr{G}_{b_3 c_3} = G_{b_3} + G_{b_3} \sum_{d_3 \subset b_2} \bar{\delta}_{d_3 b_3} V_{d_3} \sum_{c_3 \subset b_2} \mathscr{G}_{d_3 c_3} . \tag{4.39}$$

This is indeed the resolvent identity for G_{b_2} if the sum over c_3 is independent of b_3 as claimed in (4.38). To show that, we operate on (4.39) with $(1 - G_0 V_{b_3})$ from the left and get

$$(1 - G_0 V_{b_3}) \sum_{c_3 \subset b_2} \mathscr{G}_{b_3 c_3} = G_0 + G_0 \sum_{d_3 \subset b_2} \bar{\delta}_{d_3 b_3} V_{d_3} \sum_{c_3 \subset b_2} \mathscr{G}_{d_3 c_3} \tag{4.40}$$

or

$$\sum_{c_3 \subset b_2} \mathscr{G}_{b_3 c_3} = G_0 + G_0 \sum_{d_3 \subset b_2} V_{d_3} \sum_{c_3 \subset b_2} \mathscr{G}_{d_3 c_3} . \tag{4.41}$$

The rhs indeed shows no dependence on b_3. It is easy to verify that

$$\mathscr{G}_{b_3 c_3} = \delta_{b_3 c_3} G_0 + G_{b_2} V_{c_3} G_0 \tag{4.42}$$

fulfills both (4.37) and (4.38).

Now let us come back to the set (4.35). The resolvent operator acts on an expression which does not depend on b_3 if the physical solution $|\Psi_{a_2}^{c_2 c_3}\rangle \equiv |\Psi_{a_2}^{(+)}\rangle$ is inserted. Therefore we may replace G_{b_2} by (4.38) and get

$$|\Psi_{a_2}^{b_2 b_3}\rangle = |\phi_{a_2}^{b_2 b_3}\rangle + \sum_{c_3 \subset b_2} \mathscr{G}_{b_3 c_3}^{b_2} \sum_{c_2 \subset c_3} \bar{\delta}_{b_2 c_2} \sum_{d_3 \subset c_2} \bar{\delta}_{d_3 c_3} V_{d_3} |\Psi_{a_2}^{c_2 d_3}\rangle$$

$$= |\phi_{a_2}^{b_2 b_3}\rangle + \sum_{c_3 \subset b_2} \mathscr{G}_{b_3 c_3}^{b_2} \sum_{c_2 \supset c_3} \bar{\delta}_{b_2 c_2} \sum_{d_3 \subset c_2} \mathscr{V}_{c_3 d_3}^{c_2} |\Psi_{a_2}^{c_2 d_3}\rangle . \tag{4.43}$$

We have added obvious superscripts b_2 to $\mathscr{G}$ and c_2 to $\mathscr{V}$. Also we have used the fact that the relations (4.37, 38), and (4.42) remain valid for the motion of two, uncoupled, interacting pairs (the second type of two-body fragmentation). The eighteen coupled equations (4.43) now have all the desired properties.

As in the three-body case we may give them the form of multi-channel Lippmann-Schwinger equations

$$|\Psi_{a_2}^{b_2 b_3}\rangle = |\phi_{a_2}^{b_2 b_3}\rangle + \sum_{c_2 d_2} \sum_{c_3 d_3} \mathscr{G}_{0, b_3 c_3}^{b_2 c_2} \mathscr{V}_{c_3 d_3}^{c_2 d_2} |\Psi_{a_2}^{d_2 d_3}\rangle \tag{4.44}$$

with

$$\mathscr{G}_{0, b_3 c_3}^{b_2 c_2} = \delta_{b_2 c_2} \mathscr{G}_{b_3 c_3}^{b_2} \tag{4.45}$$

and

$$\mathscr{V}_{c_3 d_3}^{c_2 d_2} = \bar{\delta}_{c_2 d_2} \mathscr{V}_{c_3 d_3}^{d_2} \quad \text{(if } c_3 \subset c_2, \text{ zero otherwise)} . \tag{4.46}$$

The structure embodied in (4.44 – 46) is very important and allows the systematic extension to N particles in an inductive manner. Equations (4.45, 46) become the new $\mathscr{G}_0$ and $\mathscr{V}$, respectively, and can be used to define through a resolvent identity a new $\mathscr{G}$, which will be related to the four-body resolvent operator in a manner analogous to (4.38). We shall not follow this development further, but refer the interested readers to the very clear presentation [4.8, 6] by *Sandhas*.

We shall now demonstrate [4.8, 6] that the new set (4.44) of eighteen coupled equations has exactly one solution, namely the physical state. Again it is very important to recognize the formal structure of (4.44), which we write in obvious matrix notation with respect to the pair indices:

$$\underset{\sim}{\psi}^{b_2} = \phi^{b_2} + \mathscr{G}^{b_2} \sum_{c_2 \neq b_2} \mathscr{V}^{c_2} \underset{\sim}{\psi}^{c_2}. \tag{4.47}$$

Now we can proceed as for the set (4.26) and operate by $(\mathbb{1} - \mathscr{G}_0 \mathscr{V}^{b_2})$ from the left:

$$(\mathbb{1} - \mathscr{G}_0 \mathscr{V}^{b_2}) \underset{\sim}{\psi}^{b_2} = (\mathbb{1} - \mathscr{G}_0 \mathscr{V}^{b_2}) \phi^{b_2} + (\mathbb{1} - \mathscr{G}_0 \mathscr{V}^{b_2}) \mathscr{G}^{b_2} \sum_{c_2 \neq b_2} \mathscr{V}^{c_2} \underset{\sim}{\psi}^{c_2}. \tag{4.48}$$

The first term on the rhs is zero, since it is the homogeneous bound state formulation for the cluster(s) in the two-fragment channels. The kernel part simplifies according to (4.37). Thus we arrive at

$$\underset{\sim}{\psi}^{b_2} = \mathscr{G}_0 \sum_{c_2} \mathscr{V}^{c_2} \underset{\sim}{\psi}^{c_2}, \tag{4.49}$$

which shows explicitly the independence of $\underset{\sim}{\psi}$ on b_2. Dropping therefore the b_2, c_2 indices, we are left with

$$\begin{aligned}
\Psi_{a_2}^{b_3} &= G_{b_3} \sum_{c_2 \supset b_3} \sum_{c_3 \subset c_2} \bar{\delta}_{c_3 b_3} V_{c_3} \Psi_{c_2}^{c_3} \\
&= G_{b_3} \sum_{c_3 \neq b_3} V_{c_3} \Psi_{a_2}^{c_3}. \tag{4.50}
\end{aligned}$$

The final step is obvious. We operate by $(1 - G_0 V_{b_3})$ from the left and get

$$|\Psi_{a_2}^{b_3}\rangle = G_0 \sum_{c_3} V_{c_3} |\Psi_{a_2}^{c_3}\rangle, \tag{4.51}$$

which shows the independence of $|\Psi_{a_2}^{b_3}\rangle$ on the three-fragment index b_3. However, if there is no dependence on the auxiliary indices the system (4.43) is identical to the underlying set of Lippmann-Schwinger equations (4.15), which define the physical state $|\Psi_{a_2}^{(+)}\rangle$ uniquely.

We have now found two systems of eighteen coupled equations, (4.35) and (4.43). Whereas (4.43) has only one solution, $|\Psi_{a_2}^{b_2 b_3}\rangle \equiv |\Psi_{a_2}^{(+)}\rangle$, this desirable property has not been shown to be true for the other set. In fact, it is not true in general. Though the set (4.35) appears to be simpler, since the subsystem resolvent operators G_{b_2} show up directly, which immediately allow an approximation in the case that tightly bound clusters exist, the set is not in unique correspondence to the underlying Lippmann-Schwinger equations. This can be seen in a simplified four particle system where only $V_{12} \neq 0$, $V_{23} \neq 0$, and $V_{34} \neq 0$. In this case the set of eighteen coupled equations (4.35) splits into two groups of three coupled equations for altogether six unknowns $|\Psi^{b_2 b_3}\rangle$ and the remaining unknowns are determined by quadrature. On the other hand, for the system (4.43) the same six unknowns are coupled with each other (again the remaining ones are determined by quadrature). From the above proof, we know that the six coupled equations have the unique solution $|\Psi^{b_2 b_3}\rangle \equiv |\Psi_{a_2}^{(+)}\rangle$. It would be surprising if, in the other formulation, two uncoupled systems of three equations would also yield exactly only one type of solution, $|\Psi^{b_2 b_3}\rangle \equiv |\Psi_{a_2}^{(+)}\rangle$. One can show [4.9] that the two formulations are related by a nontrivial matrix multiplier, which in general allows for additional spurious solutions in the case (4.35).

Exercise: Work out the reduced number of coupled equations, mentioned above, and verify the existence of the matrix multiplier. This is very helpful to grasp the structure of the coupling scheme.

As for all formulations which are linked by a multiplier to a formulation which is in unique correspondence to the Schrödinger equation, the question whether additional nonphysical solutions may invalidate numerical outputs can only be safely answered through thorough numerical studies, which have not yet been comprehensively undertaken.

Once one has accepted the necessity of living with eighteen equations for four particles, one would like to know at least that the formulation is connected after a finite number of iterations. This is the case as we shall now demonstrate. First of all we recognize that the kernel in (4.35) or (4.43) is only two-body connected. The most weakly connected part has the form

$$G_0 \bar{\delta}_{b_2 c_2} \bar{\delta}_{b_3 c_3} V_{c_3} \tag{4.52}$$

with the index structure

$$b_3 \subset b_2, \quad b_3, c_3 \subset c_2, \quad b_3 \neq c_3, \quad b_2 \neq c_2. \tag{4.53}$$

In order to get connectivity for four particles, that is a sequence of three different pair interactions, at least two iterations are required. For two iterations we encounter the following sequence of indices

$$\begin{array}{c} b_2 \neq c_2 \neq d_2 \neq e_2 \\ \diagup\ |\diagup\ |\diagup\ | \\ b_3 \neq c_3 \neq d_3 \neq e_3 \,. \end{array} \tag{4.54}$$

The lines indicate the $\subset$ relations ($b_3 \subset c_2$, $c_3 \subset c_2$, $c_3 \subset d_2, \ldots$). The claim is that $e_3 \neq c_3$ which leads to a sequence of three different pair interactions $V_{c_3} G_0 V_{d_3} G_0 V_{e_3}$. This is the case, since the pairs c_3 and d_3 ($c_3 \neq d_3$) belong to a unique c_2 and can therefore not belong to another $d_2 \neq c_2$.

For scattering problems the transition operators are of central interest and not the wave function. In the two-body system, the equation for the t-operator defined by $t\,|\phi\rangle \equiv V\,|\Psi^{(+)}\rangle$ follows directly from the Lippmann-Schwinger equation:

$$t\,|\phi\rangle = V\,|\phi\rangle + V G_0 t\,|\phi\rangle .\tag{4.55}$$

Now in the four-body system, the set (4.44) has the same Lippmann-Schwinger type structure and one may define in a matrix notation "transition operators" $\mathcal{T}$:

$$\mathcal{T}\,\phi_{a_2} \equiv \mathcal{V}\,\Psi_{a_2} .\tag{4.56}$$

First we note that the matrix elements $\mathcal{T}$ yield the physical transition amplitudes:

$$(\phi^T_{b_2}|\mathcal{T}|\phi_{a_2}) = (\phi^T_{b_2}|\mathcal{V}|\Psi_{a_2}) = \sum_{k_2 k_3}\sum_{l_2 l_3} (\phi^{k_2 k_3}_{b_2}|\mathcal{V}^{k_2 l_2}_{k_3 l_3}|\Psi^{l_2 l_3}_{a_2}) .\tag{4.57}$$

The upper indices for $|\phi\rangle$ and $|\Psi\rangle$ are purely artificial and using the definition of $\mathcal{V}$ we get indeed

$$(\phi^T_{b_2}|\mathcal{T}|\phi_{a_2}) = (\phi_{b_2}|V^{b_2}|\Psi^{(+)}_{a_2}) .\tag{4.58}$$

Secondly, as in the two-body case, the set (4.44) provides immediately the matrix Lippmann-Schwinger equations for the "transition operators"

$$\mathcal{T}^{b_2 c_2}_{b_3 c_3} = \mathcal{V}^{b_2 c_2}_{b_3 c_3} + \sum_{\substack{d_2 e_2 \\ d_3 e_3}} \mathcal{V}^{b_2 d_2}_{b_3 d_3} \mathcal{G}^{d_2 e_2}_{d_3 e_3} \mathcal{T}^{e_2 c_2}_{e_3 c_3} .\tag{4.59}$$

This formulation and its generalisation to N particles has been presented in [4.8, 6].

Exercise: Establish in the same manner Lippmann-Schwinger matrix equations for $\mathcal{T}$-operators in the three-body system based on (4.26). How are they related to the AGS-equations (3.164)?

4.3 Yakubovsky Equations

For three particles, Faddeev's equations couple components of the wave function. Is there a corresponding decomposition of the four-body wave function?

The answer is yes as *Yakubovsky* [4.1] showed even for an arbitrary number of particles. We follow the derivation presented in [4.10]. The scattering state $|\Psi\rangle \equiv |\Psi_{\bar{a}_2}^{(+)}\rangle$ for instance, obeys

$$|\Psi\rangle = G_0 \sum_{a_3} V_{a_3} |\Psi\rangle , \qquad (4.60)$$

which follows from (4.10) and the obvious Lippmann identity

$$\lim_{\varepsilon \to 0} i\varepsilon \, G_0(E + i\varepsilon) \, \phi_{\bar{a}_2} = 0 .$$

As in the three-body system, this suggests a decomposition of Ψ, now into six parts:

$$|\Psi\rangle = \sum_{a_3} |\psi_{a_3}\rangle \qquad (4.61)$$

with

$$|\psi_{a_3}\rangle \equiv G_0 V_{a_3} |\Psi\rangle . \qquad (4.62)$$

Let us go back for a moment to the three-body system. In that case the iteration of (4.60) yields connected and disconnected diagrams, where in the latter type only one pair interaction operates consecutively. They are clearly generated by the term $|\psi_{a_3}\rangle$ in the decomposition of $|\Psi\rangle$ on the rhs of (4.62). That subset of disconnected diagrams was summed up in the following manner:

$$(1 - G_0 V_{a_3}) |\psi_{a_3}\rangle = G_0 V_{a_3} \sum_{b_3 \neq a_3} |\psi_{b_3}\rangle \qquad (4.63)$$

or

$$|\psi_{a_3}\rangle = G_0 t_{a_3} \sum_{b_3 \neq a_3} |\psi_{b_3}\rangle + |\overset{0}{\psi}_{a_3}\rangle . \qquad (4.64)$$

Thereby the use of the two-body t-operator allowed us to invert (4.63) through

$$(1 + G_0 t_{a_3})(1 - G_0 V_{a_3}) = 1 . \qquad (4.65)$$

The free term $|\overset{0}{\psi}_{a_3}\rangle$ is a solution of the lhs alone and its presence or absence depends on the chosen boundary conditions. For three particles, the first iteration of (4.64) yields necessarily connected diagrams, since the interactions within two-body clusters are already summed. It is plausible that this first step also has to be done for four particles and we take (4.64) as the starting point. Since for four particles, in the case that the initial channel is of the two-fragment type, there is no $|\overset{0}{\psi}_{a_3}\rangle$ which could contribute to the incoming flux, the free term is absent. Now any number of iterations of (4.64) will produce not only connected diagrams, but also the types of disconnected diagrams shown in Fig. 4.2. In the first case there is a noninteracting spectator particle and in

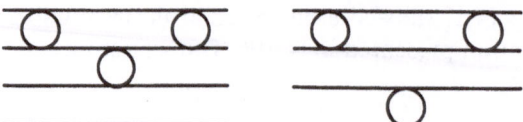

Fig. 4.2. The two types of disconnected diagrams in a four-body system once two-body t-matrices have been introduced

the second case the two pairs do not interact. This leads to δ-functions for conserved relative momenta and prohibits that any power of the kernel will be of the Fredholm type.

Exactly as in the three-body system we have to group together on the rhs of (4.64), those terms which are responsible for the occurrence of the disconnected diagrams. Since $|\psi_{b_3}\rangle$ is proportional to t_{b_3}, we select those pairs b_3 which are internal to the noninteracting subcluster of the four-body system. Together with a_3 this determines uniquely the two-body fragmentation, as we saw in the previous chapter. Let us regard an example.

$$|\psi_{12}\rangle = G_0 t_{12}(|\psi_{13}\rangle + |\psi_{14}\rangle + |\psi_{23}\rangle + |\psi_{24}\rangle + |\psi_{34}\rangle)$$

$$= G_0 t_{12}(|\psi_{24}\rangle + |\psi_{14}\rangle) + G_0 t_{12}(|\psi_{13}\rangle + |\psi_{23}\rangle) + G_0 t_{12}|\psi_{34}\rangle$$

$$\equiv |\psi_{1\,2\,4\,3;1\,2}\rangle + |\psi_{1\,2\,3,4;1\,2}\rangle + |\psi_{1\,2,3\,4;1\,2}\rangle\,. \tag{4.66}$$

Therefore in general we can decompose the rhs of (4.64) as

$$|\psi_{a_3}\rangle \equiv \sum_{a_2 \supset a_3} |\psi_{a_2 a_3}\rangle \tag{4.67}$$

with

$$|\psi_{a_2 a_3}\rangle = G_0 t_{a_3} \sum_{\substack{b_3 \neq a_3 \\ b_3 \subset a_2}} |\psi_{b_3}\rangle \tag{4.68}$$

or

$$|\psi_{a_2 a_3}\rangle = G_0 t_{a_3} \sum_{\substack{b_3 \neq a_3 \\ b_3 \subset a_2}} G_0 V_{b_3} |\Psi\rangle\,. \tag{4.69}$$

In this manner $|\Psi\rangle$ is split into eighteen components (4.69), the Yakubovsky components:

$$|\Psi\rangle = \sum_{a_3} |\psi_{a_3}\rangle = \sum_{a_3}\sum_{a_2 \supset a_3} |\psi_{a_2 a_3}\rangle = \sum_{a_2}\sum_{a_3 \subset a_2} |\psi_{a_2 a_3}\rangle\,. \tag{4.70}$$

Now we follow the strategy learned in the three-body system and collect the parts belonging to the subclusters a_2:

$$|\psi_{a_2 a_3}\rangle = G_0 t_{a_3} \sum_{\substack{b_3 \neq a_3 \\ b_3 \subset a_2}} |\psi_{b_3}\rangle = G_0 t_{a_3} \sum_{\substack{b_3 \neq a_3 \\ b_3 \subset a_2}}\sum_{b_2 \supset b_3} |\psi_{b_2 b_3}\rangle$$

$$= G_0 t_{a_3} \sum_{\substack{b_3 \neq a_3 \\ b_3 \subset a_2}} |\psi_{a_2 b_3}\rangle + G_0 t_{a_3} \sum_{\substack{b_3 \neq a_3 \\ b_3 \subset a_2}}\sum_{\substack{b_2 \supset b_3 \\ b_2 \neq a_2}} |\psi_{b_2 b_3}\rangle \tag{4.71}$$

or

$$|\psi_{a_2a_3}\rangle - G_0 t_{a_3} \sum_{\substack{b_3 \neq a_3 \\ b_3 \subset a_2}} |\psi_{a_2b_3}\rangle = G_0 t_{a_3} \sum_{\substack{b_3 \neq a_3 \\ b_3 \subset a_2}} \sum_{\substack{b_2 \supset b_3 \\ b_2 \neq a_2}} |\psi_{b_2b_3}\rangle . \tag{4.72}$$

In order to understand the necessary inversion it is very helpful [4.11] to consider an example. Let us choose $a_2 = 1\,2\,3,4$, to which belong $a_3 = 1\,2,\,1\,3$, and $2\,3$. Then we get explicitly:

$$|\psi_{a_2,12}\rangle - G_0 t_{12}(|\psi_{a_2,13}\rangle + |\psi_{a_2,23}\rangle) = G_0 t_{12} \sum_{\substack{b_3 \neq 12 \\ b_3 \subset a_2}} \sum_{b_2 \supset b_3} \bar{\delta}_{b_2 a_2}|\psi_{b_2b_3}\rangle$$

$$|\psi_{a_2,13}\rangle - G_0 t_{13}(|\psi_{a_2,12}\rangle + |\psi_{a_2,23}\rangle) = G_0 t_{12} \sum_{\substack{b_3 \neq 13 \\ b_3 \subset a_2}} \sum_{b_2 \supset b_3} \bar{\delta}_{b_2 a_2}|\psi_{b_2b_3}\rangle$$

$$|\psi_{a_2,23}\rangle - G_0 t_{23}(|\psi_{a_2,12}\rangle + |\psi_{a_2,13}\rangle) = G_0 t_{23} \sum_{\substack{b_3 \neq 23 \\ b_3 \subset a_2}} \sum_{b_2 \supset b_3} \bar{\delta}_{b_2 a_2}|\psi_{b_2b_3}\rangle \tag{4.73}$$

or in matrix notation

$$\begin{pmatrix} |\psi_{a_2,12}\rangle \\ |\psi_{a_2,13}\rangle \\ |\psi_{a_2,23}\rangle \end{pmatrix} = G_0 \begin{pmatrix} 0 & t_{12} & t_{12} \\ t_{13} & 0 & t_{13} \\ t_{23} & t_{23} & 0 \end{pmatrix} \begin{pmatrix} |\psi_{a_2,12}\rangle \\ |\psi_{a_2,13}\rangle \\ |\psi_{a_2,23}\rangle \end{pmatrix}$$

$$= G_0 \begin{pmatrix} 0 & t_{12} & t_{12} \\ t_{13} & 0 & t_{13} \\ t_{23} & t_{23} & 0 \end{pmatrix} \begin{pmatrix} \sum_{b_2 \supset 12} \bar{\delta}_{b_2 a_2}|\psi_{b_2,12}\rangle \\ \sum_{b_2 \supset 13} \bar{\delta}_{b_2 a_2}|\psi_{b_2,13}\rangle \\ \sum_{b_2 \supset 23} \bar{\delta}_{b_2 a_2}|\psi_{b_2,23}\rangle \end{pmatrix} . \tag{4.74}$$

The lhs is the now familiar Faddeev form of the three-body problem. For $a_2 = 1\,2,3\,4$ we would encounter on the lhs a problem of interacting pairs $1\,2$ and $3\,4$ but without interpair interaction.

In obvious matrix notation (4.74) reads

$$(\mathbb{1} - G_0 \underset{\approx}{t}) \psi_{a_2} = G_0 \underset{\approx}{t} R . \tag{4.75}$$

Analogously to (4.65) we introduce an inverse operator through

$$(\mathbb{1} + G_0 \underset{\approx}{K})(\mathbb{1} - G_0 \underset{\approx}{t}) = \mathbb{1} \tag{4.76}$$

and get

$$\psi_{a_2} = \overset{0}{\psi}_{a_2} + (\mathbb{1} + G_0 \underset{\approx}{K}) G_0 \underset{\approx}{t} R . \tag{4.77}$$

Equation (4.76) is equivalent to

$$K = t + K G_0 t,$$ (4.78)

which allows us to simplify the rhs of (4.77) to

$$\psi_{a_2} = \overset{0}{\psi}_{a_2} + G_0 K R .$$ (4.79)

The driving term is a solution to the lhs alone:

$$(1 - G_0 t) \overset{0}{\psi}_{a_2} = 0 .$$ (4.80)

In the above example, particle 4 is a spectator and therefore $\overset{0}{\psi}_{a_2}$ is a product of the momentum eigenstate $|q_{a_2}\rangle$ of that particle and a part depending on the remaining particles 1, 2, and 3. That last piece obviously consists of the Faddeev components $|\chi_{a_3}\rangle$ of a three-body bound state. Thus the three components

$$|\overset{0}{\psi}_{a_2 a_3}\rangle = |\chi_{a_3}\rangle |q_{a_2}\rangle$$ (4.81)

sum up to the channel state

$$|\phi_{a_2}\rangle = \sum_{a_3 \subset a_2} |\overset{0}{\psi}_{a_2 a_3}\rangle .$$ (4.82)

For $a_2 = 1\,2, 3\,4$ the homogeneous problem (4.80) is explicitly

$$|\overset{0}{\psi}_{a_2, 1\,2}\rangle - G_0 t_{1\,2} |\overset{0}{\psi}_{a_2, 3\,4}\rangle = 0$$
$$|\overset{0}{\psi}_{a_2, 3\,4}\rangle - G_0 t_{3\,4} |\overset{0}{\psi}_{a_2, 1\,2}\rangle = 0 ,$$ (4.83)

which is solved by

$$|\overset{0}{\psi}_{a_2, 1\,2}\rangle = G_0 V_{1\,2} |\varphi_{1\,2}\rangle |\varphi_{3\,4}\rangle |q\rangle$$ (4.84)

$$|\overset{0}{\psi}_{a_2, 3\,4}\rangle = G_0 V_{3\,4} |\varphi_{1\,2}\rangle |\varphi_{3\,4}\rangle |q\rangle .$$ (4.85)

The $|\varphi\rangle$'s are two-body bound states and $|q\rangle$ the momentum eigenstate of relative motion of the two, bound pairs $1\,2$ and $3\,4$. Again the two components sum up to a channel state:

$$|\overset{0}{\psi}_{a_2, 1\,2}\rangle + |\overset{0}{\psi}_{a_2, 3\,4}\rangle = G_0 (V_{1\,2} + V_{3\,4}) |\varphi_{1\,2}\rangle |\varphi_{3\,4}\rangle |q\rangle$$
$$= |\varphi_{1\,2}\rangle |\varphi_{3\,4}\rangle |q\rangle \equiv |\phi_{a_2}\rangle .$$ (4.86)

The total state $|\Psi\rangle$ contains $|\phi_{a_2}\rangle$ only in the initial channel. Therefore the driving term $\overset{0}{\psi}_{a_2}$ shows up only if a_2 is the initial two-fragment channel index $\bar{a}_2$. In explicit notation (4.79) reads

$$|\psi_{a_2 a_3}\rangle = |\overset{0}{\psi}_{a_2 a_3}\rangle \delta_{a_2 \bar{a}_2} + \sum_{b_3 \subset a_2} G_0 K^{a_2}_{a_3 b_3} \sum_{b_2 \supset b_3} \bar{\delta}_{b_2 a_2} |\psi_{b_2 b_3}\rangle .$$ (4.87)

This is the set of eighteen coupled Yakubovsky equations.

Just as the three-body Faddeev equations contained two-body t-operators in the kernel, the four-body equations (4.87) contain now three-body operators. Since K^{a_2} sums up the pair interactions within the clusters of a_2, the kernel has to get connected after iterating. Now K, as given in (4.78), contains a part which is only two-body connected, and therefore exactly two iterations are necessary. Comparing with (4.35) or (4.44), we recognize the same type of coupling scheme which reconfirms the connectivity after two iterations.

One may ask for the link between the Yakubovsky equations and the set of seven Lippmann-Schwinger equations. The Yakubovsky components are defined in (4.69). The operation which reduces $|\Psi\rangle$ to a component can be applied from the left onto all seven equations:

$$|\psi_{a_2 a_3}\rangle = G_0 t_{a_3} \sum_{b_3 \subset a_2} \bar{\delta}_{b_3 a_3} G_0 V_{b_3} |\phi_{\bar{a}_2}\rangle \delta_{a_2 \bar{a}_2} + G_0 t_{a_3} \sum_{b_3 \subset a_2} \bar{\delta}_{b_3 c_3} G_0 V_{b_3} G_{a_2} V^{a_2} |\Psi\rangle$$ (4.88)

The driving term is easy to handle. For $a_2 = \bar{a}_2$ we get

$$G_0 t_{a_3} \sum_{b_3 \subset \bar{a}_2} \bar{\delta}_{b_3 a_3} G_0 V_{b_3} |\phi_{\bar{a}_2}\rangle = \sum_{b_3 \subset \bar{a}_2} \bar{\delta}_{b_3 a_3} G_{a_3} V_{a_3} G_0 V_{b_3} |\phi_{\bar{a}_2}\rangle$$

$$= G_0 V_{a_3} \sum_{b_3 \subset \bar{a}_2} G_{a_3} V_{b_3} \bar{\delta}_{b_3 a_3} |\phi_{\bar{a}_2}\rangle$$

$$= G_0 V_{a_3} |\phi_{\bar{a}_2}\rangle \equiv |\overset{0}{\psi}_{\bar{a}_2 a_3}\rangle .$$ (4.89)

Thus, as expected, we get the Faddeev components of the subcluster which are the driving terms in (4.87).

The important point is that now the second term on the rhs of (4.87) can be rewritten in terms of Yakubovsky components:

$$G_0 t_{a_3} \sum_{b_3 \subset a_2} \bar{\delta}_{b_3 a_3} G_0 V_{b_3} G_{a_2} V^{a_2} |\Psi\rangle$$

$$= G_0 t_{a_3} \sum_{b_3 \subset a_2} \bar{\delta}_{b_3 a_3} G_0 V_{b_3} \sum_{c_3 \subset a_2} (G_0 \delta_{b_3 c_3} + G_{a_2} V_{c_3} G_0) V^{a_2} |\Psi\rangle$$

$$= G_0 t_{a_3} \sum_{b_3 \subset a_2} \bar{\delta}_{b_3 a_3} G_0 \sum_{c_3 \subset a_2} (V_{b_3} \delta_{b_3 c_3} + V_{b_3} G_{a_2} V_{c_3}) G_0 V^{a_2} |\Psi\rangle .$$ (4.90)

The expression in the bracket refers either to a three-body problem or two, uncorrelated, two-body problems. It is easy to verify that

$$M^{a_2}_{b_3 c_3} \equiv V_{b_3} \delta_{b_3 c_3} + V_{b_3} G_{a_2} V_{c_3}$$ (4.91)

obeys the Faddeev equations

$$M_{a_3c_3}^{a_2} = t_{a_3}\delta_{a_3c_3} + \sum_{\substack{b_3 \neq a_3 \\ b_3 \subset a_2}} t_{a_3} G_0 M_{b_3c_3}^{a_2} = t_{a_3}\delta_{a_3c_3} + \sum_{\substack{b_3 \neq c_3 \\ b_3 \subset a_2}} M_{a_3b_3}^{a_2} G_0 t_{c_3} . \tag{4.92}$$

Exercise: With the aid of resolvent identities verify (4.92).

Therefore the rhs of (4.90) can be written as

$$G_0 t_{a_3} \sum_{b_3 \subset a_2} \bar{\delta}_{b_3 a_3} G_0 \sum_{c_3 \subset a_2} M_{b_3c_3}^{a_2} G_0 V^{a_2} |\Psi\rangle$$

$$= G_0 \sum_{c_3 \subset a_2} \sum_{b_3 \subset a_2} \bar{\delta}_{b_3 c_3} M_{a_3b_3}^{a_2} G_0 t_{c_3} G_0 V^{a_2} |\Psi\rangle$$

$$= G_0 \sum_{b_3 \subset a_2} M_{a_3b_3}^{a_2} \sum_{c_3 \subset a_2} \bar{\delta}_{c_3 b_3} G_0 t_{c_3} G_0 \sum_{d_3 \not\subset a_2} V_{d_3} |\Psi\rangle$$

$$= G_0 \sum_{b_3 \subset a_2} M_{a_3b_3}^{a_2} \sum_{c_3 \subset a_2} \bar{\delta}_{c_3 b_3} G_0 t_{c_3} G_0 \sum_{c_2 \supset c_3} \bar{\delta}_{c_2 a_2} \sum_{\substack{d_3 \subset c_2 \\ d_3 \neq c_3}} V_{d_3} |\Psi\rangle$$

$$= G_0 \sum_{b_3 \subset a_2} M_{a_3b_3}^{a_2} \sum_{c_3 \subset a_2} \bar{\delta}_{c_3 b_3} \sum_{c_2 \supset c_3} \bar{\delta}_{c_2 a_2} |\Psi_{c_2 c_3}\rangle . \tag{4.93}$$

We used (4.33) and (4.69) in the last two steps and end up with

$$|\Psi_{a_2 a_3}\rangle = |\overset{0}{\Psi}_{\bar{a}_2 a_3}\rangle \delta_{a_2 \bar{a}_2} + \sum_{b_3 \subset a_2} G_0 M_{a_3b_3}^{a_2} \sum_{c_3 \subset a_2} \bar{\delta}_{c_3 b_3} \sum_{c_2 \supset c_3} \bar{\delta}_{c_2 a_2} |\Psi_{c_2 c_3}\rangle . \tag{4.94}$$

These are eighteen coupled equations for the Yakubovsky components and are identical to the set (4.87), as follows from the connection

$$K_{a_3c_3}^{a_2} = \sum_{b_3 \subset a_2} \bar{\delta}_{b_3 c_3} M_{a_3b_3}^{a_2} . \tag{4.95}$$

This is a simple consequence of (4.78) and (4.92).

Exercise: Verify (4.95).

Finally we regard the case of identical particles. According to the two types of two-body fragmentations, the Yakubovsky components split into two groups. There are twelve of the type $i(jkl)$ and six of the type $(ij)(kl)$. It is obvious from their definition (4.69) that the twelve and six are identical among each other after suitable particle permutations. Therefore, there remain only two unknown functions and the eighteen equations reduce to two. Note, however, that the full complexity of the coupling scheme is still present in the permutation operators.

4.4 AGS-Equations for Transition Operators

In the three-body problem, the set of three Lippmann-Schwinger equations lends itself very naturally to the derivation of three coupled equations for the transition operators $U_{\alpha\beta}$. The set of seven Lippmann-Schwinger equations for four particles appears at first sight to be equally suited to deriving seven coupled equations for the transition operators $U^{b_2 a_2}$ between two-fragment channels. They are defined by

$$U^{b_2 a_2} |\phi_{a_2}\rangle \equiv V^{b_2} |\Psi_{a_2}^{(+)}\rangle = \sum_{b_3 \Cup b_2} V_{b_3} |\Psi_{a_2}^{(+)}\rangle \tag{4.96}$$

However, we encounter immediately an ambiguity in setting up the coupling scheme. Which of the Lippmann-Schwinger equations should be used to represent $V_{b_3} |\Psi_{a_2}^{(+)}\rangle$? In order to be quite general we shall write

$$U^{b_2 a_2} = \sum_{b_3 \Cup b_2} V_{b_3} \delta_{a_2 b_2} + \sum_{b_3 \Cup b_2} V_{b_3} \sum_{c_2 \neq b_2} W_{b_3 c_2} G_{c_2} U^{c_2 a_2} . \tag{4.97}$$

The rectangular matrix W is supposed to consist of elements which are either 0 or 1 and which define the coupling scheme. Up to now, the ansatz (4.97) exhibits explicitly only the requirement that $U^{b_2 a_2}$ does not couple to itself. Further we would like to have the properties that (4.97) is a closed set for the seven transition operators and that it should be connected after a finite number of iterations. After one iteration we encounter

$$\sum_{b_3 \Cup b_2} V_{b_3} \sum_{c_2 \neq b_2} W_{b_3 c_2} G_{c_2} \sum_{c_3 \Cup c_2} V_{c_3} \sum_{d_2 \neq c_2} W_{c_3 d_2} G_{d_2} . \tag{4.98}$$

In order to approach connectivity c_3 should be different from b_3. Let us regard an example: $b_2 = 1\,2\,3,4$, $b_3 = 1\,4$. If the external interaction for c_2 should exclude $V_{1\,4}$ the allowed c_2's are: $c_2 = (1\,2\,4)3$, $(1\,3\,4)2$, and $(1\,4)(2\,3)$. The corresponding external interactions are: $V^{c_2} = V_{1\,3} + V_{2\,3} + V_{3\,4}$, $V^{c_2} = V_{1\,2} + V_{2\,3} + V_{2\,4}$, and $V^{c_2} = V_{1\,2} + V_{1\,3} + V_{2\,4} + V_{3\,4}$. Thus all pair interactions except $V_{1\,4}$ occur. Consequently after another iteration, $V_{1\,4}$ occurs again and one can never achieve connectivity. Now one may require a more severe selection of allowed c_2's by dropping $(1\,2\,4)3$ or $(1\,3\,4)2$. One cannot drop both or $(1\,4)(2\,3)$ if one wants to keep a closed set of seven equations. This does not remedy the fact that all pair interactions $V_{c_3} \neq V_{1\,4}$ still occur. Therefore there exists no choice of W which leads to a kernel which becomes connected after a certain number of iterations. One is therefore forced to extend the coupling scheme and we shall link the derivation [4.6, 8] to the results already gained for the wave function. As a first step, let us replace the subsystem operators K in the set (4.87) by transition operators. Putting

$$K_{a_3 b_3}^{b_2} \equiv t_{a_3} G_0 U_{a_3 b_3}^{b_2} \tag{4.99}$$

in (4.78) we get

$$U_{a_3b_3}^{b_2} = G_0^{-1}\bar{\delta}_{a_3b_3} + \sum_{c_3\subset b_2} U_{a_3c_3}^{b_2} G_0 t_{c_3}\bar{\delta}_{c_3b_3}. \tag{4.100}$$

For $b_2 = i(jkl)$ these are the AGS-equations for the physical transition operators in the three-body subsystems. Note however that, they are embedded in a four-particle space and are therefore off-shell. The new operators U, for $b_2 = (ij)(kl)$, obey two coupled equations. In terms of U, the Yakubovsky equations (4.87) read

$$|\psi_{a_2a_3}^{\bar{a}_2}\rangle = |\psi_{a_2a_3}^0\rangle\delta_{a_2\bar{a}_2} + \sum_{b_3\subset a_2} G_0 t_{a_3} G_0 U_{a_3b_3}^{a_2} \sum_{b_2\supset b_3} \bar{\delta}_{b_2a_2} |\psi_{b_2b_3}^{\bar{a}_2}\rangle. \tag{4.101}$$

This set can be cast into the structure of a matrix Lippmann-Schwinger equation

$$\psi^{\bar{a}_2} = \overset{0}{\psi}{}^{\bar{a}_2} + \mathbb{G}\mathbb{V}\psi^{\bar{a}_2}, \tag{4.102}$$

where new matrix operators

$$\mathbb{G}_{0\,a_3b_3}^{a_2b_2} = G_0 t_{a_3} G_0 U_{a_3b_3}^{a_2} G_0 t_{b_3} G_0 \delta_{a_2b_2} \tag{4.103}$$

and

$$\mathbb{V}_{b_3c_3}^{b_2c_2} = (G_0 t_{b_3} G_0)^{-1}\bar{\delta}_{b_2c_2}\delta_{b_3c_3} \tag{4.104}$$

have been introduced. The dependence on the initial channel has now been indicated explicitly by a new subscript. The door is now open to introduce a matrix T-operator

$$\mathbb{V}\psi \equiv \mathbb{T}\overset{0}{\psi}, \tag{4.105}$$

which obviously obeys the equation

$$\mathbb{T} = \mathbb{V} + \mathbb{V}\mathbb{G}_0\mathbb{T}. \tag{4.106}$$

This reads in an explicit notation

$$\begin{aligned}T_{a_3b_3}^{a_2b_2} &= V_{a_3b_3}^{a_2b_2} + \sum_{c_2d_2}\sum_{c_3d_3} V_{a_3c_3}^{a_2c_2}\mathbb{G}_{0\,c_3d_3}^{c_2d_2} T_{d_3b_3}^{d_2b_2}\\ &= (G_0 t_{a_3} G_0)^{-1}\bar{\delta}_{a_2b_2}\delta_{a_3b_3} + \sum_{c_2}\bar{\delta}_{c_2a_2} U_{a_3c_3}^{c_2} G_0 t_{c_3} G_0 T_{c_3b_3}^{c_2b_2}. \end{aligned} \tag{4.107}$$

These are again eighteen coupled equations for each fixed pair of indices b_2, b_3. Are these 18×18 operators $T_{a_3b_3}^{a_2b_2}$ of any use for calculating the physical transition amplitudes? The answer is yes, and we again rely heavily on formal analogies by regarding the candidate for the physical transition amplitude:

$$\langle \overset{0}{\psi}{}^{\bar{a}_2} | \mathbb{T} | \overset{0}{\psi}{}^{\bar{b}_2} \rangle = \sum_{a_2 b_2} \sum_{a_3 b_3} \langle \overset{0}{\psi}{}^{\bar{a}_2}_{a_2 a_3} | T^{a_2 b_2}_{a_3 b_3} | \overset{0}{\psi}{}^{\bar{b}_2}_{b_2 b_3} \rangle$$

$$= \sum_{a_2 b_2} \sum_{a_3 b_3} \langle \overset{0}{\psi}{}^{\bar{a}_2}_{a_2 a_3} | V^{a_2 b_2}_{a_3 b_3} | \psi^{\bar{b}_2}_{b_2 b_3} \rangle . \tag{4.108}$$

In the second step we have used the definition (4.105) of $\mathbb{T}$. Now we insert the explicit form of the Yakubovsky components and the definition (4.104) of $\mathbb{V}$:

$$\langle \overset{0}{\psi}{}^{\bar{a}_2} | \mathbb{T} | \overset{0}{\psi}{}^{\bar{b}_2} \rangle = \sum_{a_2 b_2} \sum_{a_3 b_3} \langle \overset{0}{\psi}{}^{\bar{a}_2}_{a_2 a_3} | (G_0 t_{a_3} G_0)^{-1} \bar{\delta}_{a_2 b_2}$$

$$\times \delta_{a_3 b_3} G_0 t_{a_3} \sum_{c_3 \subset b_2} \bar{\delta}_{c_3 a_3} G_0 V_{c_3} | \Psi^{(+)}_{\bar{b}_2} \rangle$$

$$= \sum_{a_2 b_2} \sum_{a_3 b_3} \langle \overset{0}{\psi}{}^{\bar{a}_2}_{a_2 a_3} | \bar{\delta}_{a_2 b_2} \bar{\delta}_{a_3 b_3} V_{b_3} | \Psi^{(+)}_{\bar{b}_2} \rangle$$

$$= \sum_{a_3 \subset \bar{a}_2} \langle \overset{0}{\psi}{}^{\bar{a}_2}_{a_3} | \sum_{b_2 \supset a_3} \delta_{b_2 \bar{a}_2} \sum_{b_3 \subset b_2} \bar{\delta}_{b_3 a_3} V_{b_3} | \Psi^{(+)}_{\bar{b}_2} \rangle . \tag{4.109}$$

The double sum in the operator is the external interaction for b_2 as we saw in (4.33). Thus

$$\langle \overset{0}{\psi}{}^{\bar{a}_2} | \mathbb{T} | \overset{0}{\psi}{}^{\bar{b}_2} \rangle = \sum_{a_3 \subset \bar{a}_2} \langle \overset{0}{\psi}{}^{\bar{a}_2}_{a_3} | V^{\bar{a}_2} | \Psi^{(+)}_{\bar{b}_2} \rangle. \tag{4.110}$$

Finally we sum over a_3, which yields the channel state according to (4.82) and (4.86), and end up with

$$\langle \overset{0}{\psi}{}^{\bar{a}_2} | \mathbb{T} | \overset{0}{\psi}{}^{\bar{b}_2} \rangle = \langle \phi_{\bar{a}_2} | V^{\bar{a}_2} | \Psi^{(+)}_{\bar{b}_2} \rangle, \tag{4.111}$$

which is the physical transition amplitude.

The set of equations (4.107) is a very convenient starting point for introducing separable approximations for subsystem transition operators [4.6, 8].

Exercise: The Faddeev equations for the three-body wave function components can be cast into the form of a matrix Lippmann-Schwinger equation analogous to (4.102). If one defines a matrix T-operator as in (4.105) which equations result?

4.5 Remarks on Equations of Higher Connectivity

The Faddeev-Yakubovsky type kernels, considered up to now, are only two-body connected. It is natural to ask for kernels of higher connectivity with at the same time a lower number of coupled equations. For four particles, the number of pair interactions, six or the number of two-fragment channels, seven, suggest themselves. Indeed sets of six coupled equations [4.12] and seven coupled equations [4.13] have been derived. Both can be found easily by inserting two types of Lippmann-Schwinger equations into each other. (This is closely related to the method of cluster decomposition [4.14]). The scattering state $|\Psi_{a_2}^{(+)}\rangle$ obeys the two types of Lippmann-Schwinger equations

$$|\Psi_{a_2}^{(+)}\rangle - G_{b_3} \sum_{b_2 \supset b_3} V_{b_2}^{b_3} |\Psi_{a_2}^{(+)}\rangle = \lim_{\varepsilon \to 0} i\varepsilon G_{b_3} |\phi_{a_2}\rangle \tag{4.112}$$

and

$$|\Psi_{a_2}^{(+)}\rangle - G_{b_2} V^{b_2} |\Psi_{a_2}^{(+)}\rangle = \lim_{\varepsilon \to 0} i\varepsilon G_{b_2} |\phi_{a_2}\rangle. \tag{4.113}$$

In (4.112) we used the easily verified identity

$$\sum_{c_3 \neq b_3} V_{c_3} = \sum_{b_2 \supset b_3} (V_{b_2} - V_{b_3}) \equiv \sum_{b_2 \supset b_3} V_{b_2}^{b_3}. \tag{4.114}$$

We insert (4.113) into (4.112) and get

$$|\Psi_{a_2}^{(+)}\rangle - G_{b_3} \sum_{b_2 \supset b_3} V_{b_2}^{b_3} G_{b_2} V^{b_2} |\Psi_{a_2}^{(+)}\rangle$$

$$= \lim_{\varepsilon \to 0} (1 + \sum_{b_2 \supset b_3} G_{b_2} V_{b_2}^{b_3}) i\varepsilon G_{b_3} |\phi_{a_2}\rangle = \delta(b_3 \subset a_2) |\phi_{a_2}\rangle. \tag{4.115}$$

In the last step we used by now familiar Lippmann identities. Let us now introduce four-body T-operators by

$$T_{b_3} |\phi_{a_2}\rangle \equiv V_{b_3} |\Psi_{a_2}^{(+)}\rangle. \tag{4.116}$$

Obviously they can be determined through (4.115) by multiplying from the left by V_{b_3}:

$$T_{b_3} |\phi_{a_2}\rangle - V_{b_3} G_{b_3} \sum_{b_2 \supset b_3} V_{b_2}^{b_3} G_{b_2} \sum_{c_2 \not\subset b_2} T_{c_3} |\phi_{a_2}\rangle$$

$$= \delta(b_3 \subset a_2) V_{b_3} |\phi_{a_2}\rangle. \tag{4.117}$$

These are six coupled equations of the type derived in [4.12]. The kernel is three-body connected and obviously becomes fully connected after one iteration. Clearly the operators (4.116) obey another set of six coupled equations, too, which follow from (4.112) by multiplication by V_{b_3} from the left. Though that set has a noncompact kernel, its solution is in unique correspondence to the Schrödinger equation. One can show [4.15, 16] that the two sets are linked by spurious multipliers, which implies, that the homogeneous system of (4.117) allows for nonphysical solutions at a set of discrete (in general complex) energies. The trap in the derivation of (4.117) was the insertion of Lippmann-Schwinger equations into each other.

A set of seven equations for the physical transition operators

$$U^{b_2 a_2}|\phi_{a_2}\rangle \equiv \sum_{b_3 \not\subset b_2} V_{b_3}|\Psi_{a_2}^{(+)}\rangle = \sum_{b_3 \not\subset b_2} T_{b_3}|\phi_{a_2}\rangle \qquad (4.118)$$

results trivially from (4.117):

$$U^{b_2 a_2} - \sum_{b_3 \not\subset b_2} V_{b_3} G_{b_3} \sum_{c_2 \supset b_3} V_{c_3}^{b_3} G_{c_2} U^{c_2 a_2} = \bar{\delta}_{b_2 a_2}(E - H_0 - V_{a_2 b_2}). \qquad (4.119)$$

The rhs is due to

$$\sum_{b_3 \not\subset b_2} \delta(b_3 \subset a_2) V_{b_3}|\phi_{a_2}\rangle = \bar{\delta}_{b_2 a_2}(E - H_0 - V_{a_2 b_2})|\phi_{a_2}\rangle \qquad (4.120)$$

where $V_{a_2 b_2}$ is the sum of pair interactions common to a_2 and b_2. The set (4.119) are the Sloan equations [4.13], which have been generalized by *Bencze* and *Redish* [4.17] to N particles. Knowing spurious solutions to the homogeneous system related to (4.117), one can obviously construct, through linear combinations of the type (4.118), spurious solutions to the homogeneous system related to Sloan's set. Whether this spuriosity defect requires special caution in numerical studies appears to still be an open question.

References

A complete survey of the literature is not intended. Many of the references cited are representative examples

Chapter 1

1.1   N. G. de Bruijn: *Asymptotic Methods in Analysis* (North-Holland, Amsterdam 1961)
1.2   C. Möller: K. Dan. Vidensk. Selsk. Mat. Fys. Medd. **23**, (1) (1945)
1.3   J. M. Cook, J. Math. Phys. Cambridge Mass. **36**, 82 (1957); M. N. Hack: Nuovo Cimento **9**, 731 (1958); J. Kupsch, W. Sandhas: Commun. Math. Phys. **2**, 147 (1966)
1.4   M. Gell-Mann, M. L. Goldberger: Phys. Rev. **91**, 398 (1953)
1.5   R. G. Newton: *Scattering Theory of Waves and Particles* (McGraw Hill, New York 1966)
1.6   B. A. Lippmann, J. Schwinger: Phys. Rev. **79**, 469 (1950)
1.7   G. Braun: Acta Phys. Austriaca **10**, 8 (1956)
1.8   A. Messiah: *Quantum Mechanics* (North-Holland, Amsterdam 1961)
1.9   J. S. Levinger: Springer Tracts Mod. Phys. **71**, 88 (1974)
1.10 R. V. Reid: Ann. Phys. **50**, 411 (1968)
1.11 A. Bohr, B. R. Mottelson: *Nuclear Structure*, Vol. I (Benjamin, New York 1969)
1.12 R. Jost, A. Pais: Phys. Rev. **82**, 840 (1951)
1.13 S. Weinberg: Phys. Rev. **131**, 440 (1963)
1.14 F. Smithies: *Integral Equations* (Cambridge University Press, New York 1958)
1.15 M. Scadron, S. Weinberg, J. Wright: Phys. Rev. **135**, B202 (1964)

Chapter 2

2.1   M. H. McGregor, M. J. Moravcsik, H. P. Stapp: Annu. Rev. Nucl. Sci. **10**, 291 (1960)
2.2   A. Messiah: *Quantum Mechanics* (North-Holland, Amsterdam 1961)
2.3   A. Bohr, B. R. Mottelson: *Nuclear Structure*, Vol. I (Benjamin, New York 1969)
2.4   M. Simonius: In *Proc. Intern. Conf. on the Few Body Problem in Nuclear and Particle Physics, Laval 1974*, ed. by R. J. Slobodrian, C. Cujac, K. Ramavataran (Les Presses de l'Université Laval, Quebec 1975) p. 160; R. B. Blin-Stoyle: *Fundamental Interactions and the Nucleus* (North-Holland, Amsterdam 1973)
2.5   F. Coester: Phys. Rev. **89**, 619 (1953); A. R. Edmonds: *Angular Momentum in Quantum Mechanics* (Princeton University Press, Princeton, New York 1957)
2.6   L. Wolfenstein, J. Ashkin: Phys. Rev. **85**, 947 (1952)
2.7   R. H. Dalitz: Proc. Phys. Soc. London **A65**, 175 (1952); J. S. Bell, F. Mandl: Proc. Phys. Soc. London **A71**, 272 (1958)
2.8   J. Binstock, R. Bryan, A. Gersten: Phys. Lett. **48B**, 77 (1974); A. R. Neghabian, W. Glöckle: Nucl. Phys. **A319**, 364 (1979); J. Binstock, R. Bryan, A. Gersten: Ann. Phys. **133**, 355 (1981)
2.9   L. Wolfenstein: Phys. Rev. **96**, 1654 (1954); B. P. Nigam: Rev. Mod. Phys. **35**, 117 (1963); J. Binstock, R. Bryan: Phys. Rev. **D9**, 2528 (1974)
2.10 A. E. Woodruft: Ann. Phys. **7**, 65 (1959)
2.11 R. G. Newton: *Scattering Theory of Waves and Particles* (McGraw Hill, New York 1966)
2.12 M. Abromowitz, I. A. Segun: *Handbook of Mathematical Functions* (Dover, New York 1968)

2.13 H. P. Stapp, T. J. Ypsilantis, N. Metropolis: Phys. Rev. **105**, 302 (1957)
2.14 J. M. Blatt, L. C. Biedenharn: Phys. Rev. **86**, 399 (1952); Rev. Mod. Phys. **24**, 258 (1952)
2.15 R. A. Bryan: In *Few Body Systems and Nuclear Forces II*, Lecture Notes in Physics, Vol. 87, ed. by H. Zingl, M. Haftel, H. Zankel (Springer, Berlin, Heidelberg, New York 1978) p. 2
2.16 R. A. Arndt, B. J. VerWest: Report DOE/ER/05223-29
2.17 M. A. Melkanoff, J. Raynal, T. Sawada: In *Methods in Computational Physics*, Vol. 6, ed. by B. Alder, S. Fernbach, M. Rotenberg (Academic, New York 1966) p. 1
2.18 K. L. Kowalski: Phys. Rev. Lett. **15**, 798 (1965); H. P. Noyes: Phys. Rev. Lett. **15**, 538 (1965)
2.19 H. S. Wall: *Continued Fractions* (Chelsea, New York 1967); G. A. Baker: In *Advances in Theoretical Physics*, Vol. 1, ed. by K. A. Brueckner (Academic, New York 1965) p. 1
2.20 J. Nuttal: J. Math. Anal. Appl. **31**, 147 (1970)
2.21 F. Smithies: *Integral Equations* (Cambridge University Press, New York 1958)
2.22 J. L. Basdevant, D. Bessis, J. Zinn-Justin: Nuovo Cimento **60A**, 185 (1969)

Chapter 3

3.1 H. Ekstein: Phys. Rev. **101**, 880 (1956); J. M. Jauch: Helv. Phys. Acta **31**, 127 (1958); **31**, 661 (1958); J. Hunziker: In *Lectures in Theoretical Physics*, ed. by A. O. Barut, W. E. Brittin (Gordon and Breach, New York 1968) p. 1
3.2 W. Sandhas: In *Few Body Nuclear Physics*, ed. by G. Pisent, V. Vanzani, L. Fonda (IAEA, Vienna 1978) p. 1
3.3 B. A. Lippmann: Phys. Rev. **102**, 264 (1956); G. Bencze, C. Chandler: Phys. Lett. **90A**, 162 (1982)
3.4 L. L. Foldy, W. Tobocman: Phys. Rev. **105**, 1099 (1957)
3.5 L. D. Faddeev: Sov. Phys. JETP **12**, 1014 (1961)
3.6 W. Glöckle: Nucl. Phys. **A141**, 620 (1970); and in *The Nuclear Many-Body Problem*, ed. by F. Calogero et al. (Editre Composotori, Bologna 1973) p. 349
3.7 S. Mukherjee: Phys. Lett. **80B**, 73 (1978)
3.8 S. K. Adhikari, W. Glöckle: Phys. Rev. **C21**, 54 (1980)
3.9 L. Lovitch: Nuovo Cimento **68A**, 81 (1982)
3.10 E. Gerjuoy: Phys. Rev. **109**, 1806 (1957); Ann. Phys. **5**, 58 (1958)
3.11 W. Sandhas: In *Few Body Dynamics*, ed. by A. N. Mitra et al. (North-Holland, Amsterdam 1976) p. 540
3.12 W. Tobocman: Phys. Rev. **C11**, 43 (1975); G. Cattapan, V. Vanzani: Phys. Rev. **C19**, 1168 (1979)
3.13 L. D. Faddeev: *Mathematical Aspects of the Three Body Problem in Quantum Scattering Theory* (Davey, New York 1965)
3.14 B. R. Karlsson, E. M. Zeiger: Phys. Rev. **D11**, 939 (1975)
3.15 D. Eyre, T. A. Osborn: Phys. Rev. **C20**, 869 (1979)
3.16 C. Lovelace: Phys. Rev. **B135**, 1125 (1964)
3.17 R. D. Amado: In *Elementary Particle Physics and Scattering Theory, 1967 Brandeis University Summer Institute of Theoretical Physics,* ed. by M. Chretien, S. S. Schweber (Gordon and Breach, New York 1970) p. 1
3.18 W. Glöckle: Z. Phys. **271**, 31 (1974)
3.19 N. G. de Bruijn: *Asymptotic Methods in Analysis* (North-Holland, Amsterdam 1961)
3.20 C. Gignoux, A. Laverne, S. P. Merkuriev: Phys. Rev. Lett. **33**, 1350 (1974); S. P. Merkuriev, C. Gignoux, A. Laverne: Ann. Phys. **99**, 30 (1976)
3.21 J. J. Benayoun, J. Chauvin, C. Gignoux, A. Laverne: Phys. Rev. Lett. **36**, 1438 (1976); A. Laverne, C. Gignoux: Nucl. Phys. **A203**, 597 (1973)
3.22 G. L. Payne, J. L. Friar, B. F. Gibson, I. R. Afnan: Phys. Rev. **C22**, 823 (1980); G. Payne, B. F. Gibson, J. L. Friar: Phys. Rev. **C22**, 832 (1980); J. L. Friar, B. F. Gibson, G. L. Payne: Z. Phys. **A301**, 309 (1981); G. L. Payne: In *Proc. of the Ninth Intern. Conf. on the*

Few Body Problem, Eugen, Oregon 1980, ed. by F. S. Levin (North-Holland, Amsterdam 1981) p. 61c; and in Nucl. Phys. **A353**, 61c (1981)

3.23 W. Glöckle, D. Heiss: Nucl. Phys. **A122**, 343 (1968)

3.24 W. Glöckle: Nucl. Phys. **A158**, 257 (1970); W. Glöckle: Nucl. Phys. **A175**, 337 (1971); W. Glöckle, K. Ueta: Z. Phys. **258**, 64 (1973)

3.25 D. J. Kouri, F. S. Levin: Nucl. Phys. **A250**, 127 (1975); D. J. Kouri, F. S. Levin: Phys. Lett. **50B**, 421 (1974)

3.26 S. K. Adhikari, W. Glöckle: Phys. Rev. **C19**, 616 (1979)

3.27 P. Federbush: Phys. Rev. **148**, 1551 (1966)

3.28 S. Weinberg: Phys. Rev. **133**, 3232 (1964)

3.29 R. G. Newton: Phys. Rev. **153**, 1502 (1967)

3.30 C. Chandler: Nucl. Phys. **A301**, 1 (1978); K. L. Kowalski: Lett. Nuovo Cimento **22**, 531 (1978); V. Vanzani: Lett. Nuovo Cimento **23**, 586 (1978)

3.31 E. O. Alt, P. Grassberger, W. Sandhas: Nucl. Phys. **B2**, 167 (1967)

3.32 K. L. Kowalski: Phys. Rev. **D7**, 1806 (1973); Nucl. Phys. **A264**, 173 (1976); W. Glöckle, R. Brandenburg: Phys. Rev. **C27**, 83 (1983)

3.33 B. H. J. McKellar: In *Few Body Dynamics,* ed. by A. N. Mitra et al. (North-Holland, Amsterdam 1976) p. 508

3.34 D. Z. Freedman, C. Lovelace, J. M. Namyslowski: Nuovo Cimento **43A**, (N02) 258 (1966)

3.35 G. F. Chew: Phys. Rev. **80**, 196 (1950); H. Kottler, K. L. Kowalski: Phys. Rev. **138B**, 619 (1965)

3.36 R. A. Malfliet, J. A. Tjon: Nucl. Phys. **A127**, 161 (1969)

3.37 W. Glöckle: Nucl. Phys. **A381**, 343 (1982)

3.38 L. Müller: Nucl. Phys. **A360**, 331 (1981)

3.39 C. Hadjuk, P. U. Sauer: Nucl. Phys. **A369**, 321 (1981)

3.40 I. H. Sloan: Phys. Rev. **185**, 1361 (1969)

3.41 H. Miyazawa: Phys. Rev. **104**, 1741 (1956); S. A. Coon, M. D. Scadron, P. C. McNamee, B. R. Barrett, D. W. E. Blatt, B. H. J. McKellar: Nucl. Phys **A317**, 242 (1979)

3.42 S. A. Coon, W. Glöckle: Phys. Rev. **23**, 1790 (1981)

3.43 J. S. Levinger: Springer Tracts Mod. Phys. **71**, 88 (1974)

3.44 S. K. Adhikari: Phys. Rev. **C10**, 1623 (1974); I. H. Sloan, S. K. Adhikari: Nucl. Phys. **A235**, 352 (1974); S. K. Adhikari, I. H. Sloan: Nucl. Phys. **A241**, 429 (1975); **125**, 297 (1975); Phys. Rev. **C11**, 1133 (1975); **C12**, 1152 (1975)

3.45 G. G. Ohlsen, Nucl. Instrum. Methods **37**, 240 (1965)

3.46 B. Zeitnitz, R. Maschuw, P. Suhr, W. Ebenhöh: In *Few Particle Problems in the Nuclear Interaction*, ed. by I. Slaus et al. (North-Holland, Amsterdam 1972) p. 117

3.47 K. Yamaguchi: Phys. Rev. **95**, 1628 (1954)

3.48 E. W. Schmid, H. Ziegelmann: *The Quantum Mechanical Three-Body Problem,* Vieweg Tracts in Pure and Applied Physics (Pergamon, New York 1974)

3.49 W. Meier: Private communication

3.50 R. D. Amado: Phys. Rev. **132**, 485 (1963)

3.51 A. C. Phillips: Phys. Lett. **20**, 529 (1966); A. C. Philipps: Phys. Rev. **142**, 984 (1966); R. A. Aaron, R. D. Amado: Phys. Rev. **150**, 857 (1966); R. T. Cahill, I. H. Sloan: Nucl. Phys. **A165**, 161 (1971)

3.52 W. Ebenhöh: Nucl. Phys. **A191**, 97 (1972)

3.53 P. Doleshall: Phys. Lett. **38B**, 298 (1972); **40B**, 443 (1972); Nucl. Phys. **A201**, 264 (1973); **A220**, 491 (1974)

3.54 W. Grübler: In *Proc. of the Ninth Intern. Conf. Few Body Problem,* Eugen, Oregon 1980, ed. by F. S. Levin (North-Holland, Amsterdam 1981) p. 31c; and in Nucl. Phys. **A353**, 31c (1981)

3.55 W. M. Kloet, J. A. Tjon: Ann. Phys. **79**, 407 (1973)

3.56 H. Ziegelmann: Z. Naturforsch. **30A**, 434 (1975)

3.57 W. M. Kloet, J. A. Tjon: Nucl. Phys. **A210**, 380 (1973)

3.58 C. Stolk, J. A. Tjon: Nucl. Phys. **A295**, 384 (1978)

3.59 B. Charnomordic, C. Fayard, G. H. Lamot: Phys. Rev. **C15**, 864 (1977)

3.60 L. M. Delves: In *Advances in Nuclear Physics,* Vol. 5, ed. by M. Baranger, E. Vogt (Plenum Press, New York 1973) p. 1; M. A. Hennell, L. M. Delves: Phys. Lett. **40B**, 20 (1972); Nucl. Phys. **A246**, 490 (1975); L. M. Delves: In *Few Body Problems in Nuclear and Particle Physics,* ed. by R. J. Slobodrian et al. (Les Presses de l'Université Laval, Quebec, 1975) p. 446

3.61 M. R. Strayer, P. U. Sauer: Nucl. Phys. **A231**, 1 (1974)

3.62 P. Numberg, D. Prosperi, E. Pace: Nucl. Phys. **A258**, 58 (1977)

3.63 Y. A. Simonov: In *The Nuclear Many Body Problem,* ed. by F. Calogero, C. Ciofi Degli Atti (Editrice-Composotori-Bologna 1973) p. 527; J. L. Ballot, M. Fabre de la Ripelle: Ann. Phys. **127**, 62 (1980)

3.64 E. P. Harper, Y. E. Kim, A. Tubis: Phys. Rev. Lett. **28**, 1533 (1972); R. A. Brandenburg, Y. E. Kim, A. Tubis: Phys. Rev. **C12**, 1368 (1975)

3.65 I. R. Afnan, N. D. Birrell: Phys. Rev. **C16**, 823 (1977)

3.66 W. Glöckle, R. Offermann: Phys. Rev. **C16**, 2039 (1977)

3.67 K. Erkelenz: Phys. Rep. **13C**, 191 (1974); K. Holinde, R. Machleidt: Nucl. Phys. **A247**, 495 (1975)

3.68 W. Glöckle, L. Müller: Phys. Rev. **C23**, 1183 (1981)

3.69 J. M. Blatt, V. F. Weisskopf: *Theoretical Nuclear Physics* (Wiley and Sons, New York 1952)

3.70 A. R. Edmonds: *Angular Momentum in Quantum Mechanics* (Princeton University Press, New Jersey 1957)

3.71 E. P. Harper, Y. E. Kim, A. Tubis: Phys. Rev. **C2**, 877 (1970); **6**, 126 (1972); R. A. Malfliet, J. A. Tjon: Ann. Phys. **61**, 425 (1970); see also [Ref. 3.21]

3.72 W. Glöckle, G. Hasberg, A. R. Neghabian: Z. Phys. **A305**, 217 (1982)

3.73 J. Torre, J. J. Benayoun, J. Chauvin: Z. Phys. **A300**, 319 (1981)

3.74 I. R. Afnan, J. M. Read: Phys. Rev. **C12**, 293 (1975)

3.75 M. Lacombe, B. Loisseau, S. M. Richard, R. Vinh Mau, J. Côté, P. Pirès, R. de Tourreil: Phys. Rev. **C21**, 861 (1980)

3.76 R. Reid: Ann. Phys. **50**, 411 (1968)

3.77 R. A. Brandenburg, P. U. Sauer, R. Machleidt: Z. Phys. **A280**, 93 (1977)

Chapter 4

4.1 O. A. Yakubovsky: Sov. J. Nucl. Phys. **5**, 937 (1967)

4.2 P. Grassberger, W. Sandhas: Nucl. Phys. **B2**, 181 (1967)

4.3 E. O. Alt, P. Grassberger, W. Sandhas: JINR Report E4-6688(1972); and in *Few Particle Problems in the Nuclear Interaction,* ed by I. Slaus, S. A. Moszkowski, R. P. Haddock, W. T. H. van Oers (North-Holland, Amsterdam 1972) p. 299; W. Sandhas: In *Progress in Particle Physics,* ed. by P. Urban, Acta Phys. Austriaca, Suppl. **13**, 679 (1974); Czech. J. Phys. **B25**, 251 (1975)

4.4 G. Bencze, C. Chandler: Phys. Lett. **90A**, 162 (1982)

4.5 G. Cattapan, V. Vanzani: Phys. Rev. **C19**, 1168 (1979)

4.6 W. Sandhas: In *Few-Body Nuclear Physics,* ed. by G. Pisent, V. Vanzani, L. Fonda (IAEA, Vienna 1978) p. 1

4.7 L. Lovitch, V. Vanzani: Lett. Nuovo Cimento **26**, 65 (1979)

4.8 W. Sandhas: In *Few Body Dynamics,* ed. by A. N. Mitra et al. (North-Holland, Amsterdam 1976) p. 540

4.9 W. Glöckle: unpublished

4.10 B. R. Karlsson, E. M. Zeiger: Phys. Rev. **D9**, 1761 (1974); **D10**, 1291 (1974)

4.11 L. D. Faddeev: In *Three Body Problem in Nuclear and Particle Physics,* ed. by J. S. C. McKee, P. M. Rolph (North-Holland, Amsterdam 1970) p. 154

4.12 A. N. Mitra, J. Gillespie, R. Sugar, N. Panchapakesan: Phys. Rev. **140**, B1336 (1965); L. Rosenberg: Phys. Rev. **140**, B217 (1965); V. A. Alessandrini: J. Math. Phys. **7**, 215

(1966); Y. Takahaski, N. Mishima: Progr. Theor. Phys. (Kyoto) **34**, 498 (1965); N. Mishima, Y. Takahaski, Progr. Theor. Phys. (Kyoto) **35**, 440 (1966)
4.13 I. H. Sloan: Phys. Rev. **C6**, 1945 (1972)
4.14 I. M. Narodetsky, D. A. Yakubovsky: Sov. J. Nucl. Phys. **14**, 178 (1972); K. Hepp: Helv. Phys. Acta **42**, 425 (1969)
4.15 V. Vanzani: Lett. Nuovo Cimento **23**, 586 (1978)
4.16 S. K. Adhikari, W. Glöckle: Phys. Rev. **C19**, 616 (1979)
4.17 G. Bencze: Nucl. Phys. **A210**, 568 (1973); E. F. Redish: Nucl. Phys. **A225**, 16 (1974)

Reviews, Monographs, and Conferences

Reviews and Presentations of Special Topics

Duck, I.: Three-particle scattering. A review of recent work on the nonrelativistic theory. Adv. Nucl. Phys. **1**, 343 (1968)

Mitra, A. N.: The nuclear three-body problem. Adv. Nucl. Phys. **3**, 1 (1969)

Noyes, H. P.: Physical information needed to solve the three-nucleon problem. Prog. Nucl. Phys. **10**, 357 (1969)

Amado, R. D.: The three-nucleon problem. Ann. Rev. Nucl. Sci. **19**, 61 (1969)

Delves, L. M., Phillips, A. C.: Present status of the nuclear three-body problem. Rev. Mod. Phys. **41**, 497 (1969)

Amado, R. D.: In *The Three-Body Problem in Elementary Particle Physics and Scattering Theory*, Vol. 2, ed. by M. Chretien and S. S. Schweber (Brandeis, New York 1970) p. 3

McKee, J. S.: The three-body problem in nuclear physics. Rep. Prog. Phys. **33**, 691 (1970)

Delves, L. M.: Variational Techniques in the nuclear three-body problem. Adv. Nucl. Phys. **5**, 1 (1972)

Sandhas, W.: The three-body problem in elementary particle physics, ed. by P. Urban. Acta Phys. Austriaca, Suppl. **IX**, 57 (1972)

Kim, Y. E., Tubis, A.: The theory of three-nucleon systems. Ann. Rev. Nucl. Sci. **24**, 69 (1974)

Levinger, J. S.: The two- and three-body problem. Springer Tracts Mod. Phys. **71**, 88 (1974)

Sandhas, W.: The N-body problem, In *Progress in Particle Physics,* ed. by P. Urban, Acta Phys. Austr., Suppl. **XIII**, 679 (1974)

Srivastava, M. K., Sprung, D. W. L.: Off-shell behaviour of the nucleon-nucleon interaction. Adv. Nucl. Phys. **8**, 121 (1975)

Phillips, A. C.: Three-body systems in nuclear physics. Rep. Prog. Phys. **40**, 905 (1977)

Thomas, A. W. (ed.): *Modern Three-Hadron Physics,* Topics Curr. Phys., Vol. 2 (Springer, Berlin, Heidelberg, New York 1977)

Monographs on Few Bodies and Related Subjects

Faddeev, L. D.: *Mathematical Aspects of the Three-Body Problem in the Quantum Scattering Theory*. (Steklov Mathematical Institute, Leningrad 1963) [English transl.: Israel Program for Scientific Translation, Jerusalem 1965]

Schmid, E. W., Ziegelmann, H.: *The Quantum Mechanical Three-Body Problem* (Pergamon Press, Oxford 1974)

Goldberger, M. L., Watson, K. M.: *Collison Theory* (Wiley, New York 1964)

Newton, R. G.: *Scattering Theory of Waves and Particles* (McGraw-Hill, New York 1966)

Rodberg, L. S., Thaler, R. M.: *Introduction to the Quantum Theory of Scattering* (Academic Press, New York 1967)

Taylor, J. R.: *Scattering Theory* (Wiley, New York 1972)

Workshops and Conferences

Nuclear Forces and the Few Nuclear Problem (Pergamon Press, London 1959)

Few-Body Problems, Light Nuclei, and Nuclear Interactions, Brela, 1967, ed. by G. Paić, I. Slaus (Gordon and Breach, New York 1968)

Three-Particle Scattering in Quantum Mechanics, College Station, Texas 1968, ed. by J. Gillespie, J. Nuttal (Benjamin, New York 1968)

Three-Body Problems in Nuclear and Particle Physics, Birmingham 1969, ed. by J. S. C. McKee, P. M. Rolph (North-Holland, Amsterdam 1970)

The Nuclear Three-Body Problem and Related Topics, Budapest 1971, Acta Phys. Acad. Sci. Hung. **33**, 102 (1973)

Few-Particle Problems in the Nuclear Interaction, Los Angeles 1972, ed. by I. Slaus, S. A. Moszkowski, R. P. Haddock, W. T. H. van Oers (North-Holland, Amsterdam 1972)

The Nuclear Many-Body Problem, Rome 1972, ed. by F. Calogero, C. Ciofi degli Atti (Editrice Compositori, Bologna 1973)

Few-Body Problems in Nuclear and Particle Physics, Laval 1974, ed. by R. J. Slobodrian, B. Cujec, K. Ramavataram (Les Presses de l'Université, Laval 1975)

Few-Body Dynamics, Delhi 1975/76, ed. by A. N. Mitra, I. Slaus, V. S. Bhasin, V. K. Gupta (North-Holland, Amsterdam 1976)

Few-Body Nuclear Physics, Trieste 1978, ed. by G. Pisent, V. Vanzani, L. Fonda (IAEA, Vienna 1978)

Few-Body Systems and Nuclear Forces I, II, Graz 1978, ed. by H. Zingl, M. Haftel, H. Zankel, Lecture Notes Phys., Vols. 82, 87 (Springer, Berlin, Heidelberg, New York 1978)

The Few-Body Problem, Eugene 1980, ed. by F. S. Levin (North-Holland, Amsterdam 1981)

Subject Index

Text and Monographs in Physics

Editors: W. Beiglböck, M. Goldhaber, E. H. Lieb, T. Regge, W. Thirring

O. Bratteli, D. W. Robinson

Operator Algebras and Quantum Statistical Mechanics 2

Equilibrium States. Models in Quantum Statistical Mechanics
1981. XI, 505 pages. ISBN 3-540-10381-3

Contents: States in Quantum Statistical Mechanics: Introduction. Continuous Quantum Systems. I. KMS States. Stability and Equilibrium. Notes and Remarks. – Models of Quantum Statistical Mechanics: Introduction. Quantum Spin Systems. Continuous Quantum Systems. II. Conclusion. – Notes and Remarks. – References. – Books and Monographs. – Articles. – List of Symbols. – Subject Index. – Corrigenda to Volume 1.

G. Gallavotti

The Elements of Mechanics

1983. 53 figures. XIV, 575 pages. ISBN 3-540-11753-9

Contents: Phenomenic Reality and Models. – Qualitative Aspects of the One-Dimensional Motion. – Systems with Many Degrees of Freedom. Theory of the Constraints. Analytical Mechanics. – Special Mechanical Systems. – Stability Properties for Dissipative and Conservative Systems. – Appendices A–P. – Definition and Symbols. – References. – Index.

J. M. Jauch, R. Fohrlich

The Theory of Photons and Electrons

The Relativistic Quantum Field Theory of Charged Particles with Spin One-half
2nd corrected printing of the 2nd expanded edition. 1980. 64 figures, 10 tables. XIX, 533 pages. ISBN 3-540-07295-0

Contents: General Principles. – The Radiation Field. – Relativistic Theory of Free Electrons. Interaction of Radiation with Electrons. – Invariance Properties of the Coupled Fields. – Subsidiary Condition and Longitudinal Field. – The S-Matrix. – Evaluation of the S-Matrix. – The Divergences in the Iteration Solution. – Renormalization. – The Photon-Electron System. – The Electron-Electron System. – The Photon-Photon System. – Theory of the External Field. – External Field Problems. – Special Problems. – Mathematical Appendix. – Supplement for the Second Edition. – Author Index. – Subject Index.

G. Ludwig

Foundations of Quantum Mechanics I

Translated from the German by C. A. Hein
1983. XII, 426 pages. ISBN 3-540-11683-4

Contents: The Problem: An Axiomatic Basis for Quantum Mechanics. – Microsystems, Preparation, and Registration Procedures. – Ensembles and Effects. – Coexistent Effects and Coexistent Decompositions. – Transformations of Registration and Preparation Procedures. Transformations of Effects and Ensembles. – Representation of Groups by Means of Effect Automorphisms and Mixture Automorphisms. – The Galileo Group. – Composite Systems. – Appendix 1: Summary of Lattice Theory. – Appendix 2: Remarks about Topological and Uniform Structures. – Appendix 3: Banach Spaces. – Appendix 4: Operators in Hilbert Space. – References. – List of Frequently Used Symbols. – List of Axioms. – Index.

R. G. Newton

Scattering Theory of Waves and Particles

2nd edition. 1982. 35 figures. XX, 743 pages.
ISBN 3-540-10950-1
(Originally published by McGraw Hill, 1966)

Contents: Scattering of Electromagnetic Waves: Formalism and General Results. Spherically Symmetric Scatterers. Limiting Cases and Approximations. Miscellaneous. – Scattering of Classical Particles: Particle Scattering in Classical Mechanics. – Quantum Scattering Theory: Time-Dependent Formal Scattering Theory. Time-Independent Formal Scattering Theory. Cross Sections. Formal Methods of Solution and Approximations. Single-Channel Scattering (Three-Dimensional Analysis in Specific Representations). Single-Channel Scattering of Spin 0 Particles, I. Single-Channel Scattering of Spin 0 Particles, II. The Watson-Regge Method (Complex Angular Momentum). Examples. Elastic Scattering of Particles with Spin. Inelastic Scattering and Reactions (Multi-channel Theory), I. Inelastic Scattering and Reactions (Multi-channel Theory), II. Short-Wavelength Approximations. The Decay of Unstable States. The Inverse Scattering Problem. – Bibliography. – Index.

Springer-Verlag
Berlin
Heidelberg
New York
Tokyo

R.D. Richtmyer

**Principles of
Advanced Mathematical Physics II**

1981. 60 figures. XI, 322 pages. ISBN 3-540-10772-X

Contents: Elementary Group Theory. - Continuous Groups. - Group Representations I: Rotations and Spherical Harmonics. - Group Representations II: General; Rigid Motions; Bessel Functions. - Group Representations and Quantum Mechanics. - Elementary Theory of Manifolds. - Covering Manifolds. - Lie Groups. - Metric and Geodesics on a Manifold. - Riemannian, Pseudo-Riemannian, and Affinely Connected Manifolds. - The Extension of Einstein Manifolds. - Bifurcations in Hydrodynamic Stability Problems. - Invariant Manifolds in the Taylor Problem. - The Early Onset of Turbulence. - References. - Index.

P. Ring, P. Schuck

The Nuclear Many-Body Problem

1980. 171 figures. XVII, 716 pages. ISBN 3-540-09820-8

Contents: The Liquid Drop Model. - The Shell Model. - Rotation and Single-Particle Motion. - Nuclear Forces. - The Hartree-Fock Method. - Pairing Correlations and Superfluid Nuclei. - The Generalized Single-Particle Model (HFB Theory). - Harmonic Vibrations. - Boson Expansion Methods. -the Generator Coordinate Method. - Restoration of Broken Symmetries. - The Time Dependent Hartree-Fock Method (TDHF). - Semiclassical Methods in Nuclear Physics. - Appendices A-F. - Bibliography. - Author Index. - Subject Index.

R.M. Santilli

**Foundations
of Theoretical Mechanics II**

Birkhoffian Generalization of Hamiltonian Mechanics
1983. XIX, 370 pages. ISBN 3-540-09482-2

Contents: Introduction. - Birkhoff's Equations. - Transformation Theory of Birkhoff's Equations. - Generalization of Galilei's Relativity. Appendix A: Indirect Lagrangian Representations. - References. - Index.

M.D. Scadron

Advanced Quantum Theory

and Its Applications Through Feynman Diagrams
Corrected 2nd printing. 1981. 78 figures. XIV, 386 pages. ISBN 3-540-10970-6

Contents: Transformation Theory: Introduction. Transforamtions in Space. Transformations in Space-Time. Boson Wave Equations. Spin-$\frac{1}{2}$ Dirac Equation. Discrete Symmetries. - Scattering Theory: Formal Theory of Scattering. Simple Scattering Dynamics. Nonrelativistic Perturbation Theory. - Covariant Feynman Diagrams: Covariant Feynman Rules. Lowest-Order Electromagnetic Interactions. Low-Energy Strong Interactions. Lowest-Order Weak Interactions. Lowest-Order Gravitational Interactions. Higher-Order Covariant Feynman Diagrams. - Problems. - Appendices. - Bibliography. - Index.

Springer-Verlag
Berlin
Heidelberg
New York
Tokyo

F.J. Ynduráin

Quantum Chromodynamics

An Introduction to the Theory of Quarks and Gluons
1983. XI, 227 pages. ISBN 3-540-11752-0

Contents: Generalities. - QCD as a Field Theory. - Deep Inelastic Processe - Quark Masses, PCAC, Chiral Dynamics, and the QCD Vacuum. - Functional Methods, Nonperturbative Solution. - References. - Index.